BOSTON STUDIES IN THE PHILOSOPHY OF SCIENCE

VOLUME XXV

THE METHOD OF ANALYSIS

SYNTHESE LIBRARY

MONOGRAPHS ON EPISTEMOLOGY,

LOGIC, METHODOLOGY, PHILOSOPHY OF SCIENCE,

SOCIOLOGY OF SCIENCE AND OF KNOWLEDGE,

AND ON THE MATHEMATICAL METHODS OF

SOCIAL AND BEHAVIORAL SCIENCES

Editors:

ROBERT S. COHEN, *Boston University*

DONALD DAVIDSON, *Rockefeller University and Princeton University*

JAAKKO HINTIKKA, *Academy of Finland and Stanford University*

GABRIËL NUCHELMANS, *University of Leyden*

WESLEY C. SALMON, *University of Arizona*

VOLUME 75

BOSTON STUDIES IN THE PHILOSOPHY OF SCIENCE

EDITED BY ROBERT S. COHEN AND MARX W. WARTOFSKY

VOLUME XXV

JAAKKO HINTIKKA AND UNTO REMES

THE METHOD OF ANALYSIS

Its Geometrical Origin and Its General Significance

D. REIDEL PUBLISHING COMPANY

DORDRECHT-HOLLAND / BOSTON-U.S.A.

Library of Congress Catalog Card Number 74–81938

Cloth edition: ISBN 90 277 0532 1
Paperback edition: ISBN 90 277 0543 7

Published by D. Reidel Publishing Company,
P.O. Box 17, Dordrecht, Holland

Sold and distributed in the U.S.A., Canada and Mexico
by D. Reidel Publishing Company, Inc.
306 Dartmouth Street, Boston,
Mass. 02116, U.S.A.

Printed in The Netherlands by D. Reidel, Dordrecht

EDITORIAL NOTE

With this volume, we continue the publication of monographs and proceedings of significant conferences in the history and philosophy of science. The Jyväskylä conference, sponsored by the International Union for the History and Philosophy of Science and organized by an international committee under the chairmanship of Professor John Murdoch, was materially supported by the Union, the Academy of Finland and the University of Jyväskylä, and by a grant to the Boston University Center from the publisher of these *Studies*. The Jyväskylä conference, and this monograph by Professor Hintikka and Mr. Remes demonstrate that fusion of historical, epistemological and scientific understanding which we have endeavored to encourage in the *Boston Studies*.

It is a particular pleasure to note the efficient and warm-hearted work of the staff at the University of Jyväskylä and of our administrative director at the Boston University Center at that time, Mrs. Elizabeth McCoy Faught, who also served as Professor Murdoch's administrator in Jyväskylä.

Center for the Philosophy and History of Science,   R. S. COHEN
Boston University   M. W. WARTOFSKY

TABLE OF CONTENTS

FOREWORD

As official sponsors of the First International Conference in the History and Philosophy of Science, the two Divisions of the International Union of the History and Philosophy of Science owe a great deal to the University of Jyväskylä and the 1973 Jyväskylä Summer Festival for the extraordinarily generous hospitality they provided. But there is an additional debt owed, not simply for the locale but for the very substance of the Conference, to the two Finnish scholars who have jointly authored the present volume. For this volume represents not only the first part of the published proceedings of this First International Conference in the History and Philosophy of Science, but also, most fittingly, the paper that opened the Conference itself.

Yet the appropriateness of the paper from which this book has resulted opening the Conference lies far less in the fact that it was a contribution by two Finnish authors to a meeting hosted in Finland than it does to the fact that this paper, and now the present book, comes to grips in an extremely direct way with the very problem the whole Conference was from the outset designed to treat. Generally put, this problem was to bring together a number of historians and philosophers of science whose contributed papers would bear witness to the ways in which the two disciplines can be, and are, of value to each other. Understandably, the specific ways in which the history and philosophy of science have been brought to bear upon one another are too numerous to review here, even in summary fashion. On a more general level, however, it is perhaps fair to say that past collaboration of the two fields has surfaced more in the form of the history of science furnishing raw materials for the philosopher of science. This has been the case more and more in recent years as philosophers have increasingly come to see the importance of dealing with the specific notions and procedures of science as *actually done* in contrast to the practice – not infrequently to be found in much earlier philosophy of science – of tailoring such notions and procedures out of largely philosophical cloth.

One would not wish to deny that the present volume is an instance of such utilization of historical materials in elucidating much of value to the philosophy of science. Yet I believe that its primary significance as an excellent example of the successful collaboration of history and philosophy points in the other direction, a direction that is far rarer in the literature – and even, I think, in informal, *viva voce* interchanges – joining the two disciplines. It consists, in a few words, in employing one or another conception, technique, or doctrine drawn from the philosophy of science in *doing* the history of science. I would not wish to claim that this is all that the present book does, but it is surely, I would judge, among the most important of its accomplishments.

What is more, it brings the analytic tools of modern logic and philosophy to bear upon an historical issue that is of major importance. Although it may be true, as the authors point out, that the *locus classicus* of the distinction between analysis and synthesis derives from Pappus of Alexandria and that the primary field of exercise of this distinction is to be found in Greek mathematics and in the traɑition arising from it, the relevant history clearly does not stop there. With roots that perhaps go back to Plato, the conscious pairing and contrasting of the methods oɪ analysis and synthesis extend – as Chapter VIII of the present book notes – throughout much of Greek medicine and later Greek philosophy, especially that found in the Neo-platonic commentators on Aristotle. From there the trail leads through medieval and Renaissance philosophy – where analysis and synthesis appear as the Latin *resolutio* and *compositio* – to early modern science itself, the latter again appropriately noted in the final chapter of this book.

Some historians of science may feel a certain amount of uneasiness, if not alarm, at the prospect of the 'translation' or analysis of a distinction arising in ancient science in terms that derive ultimately from contemporary logic and analytic philosophy. This will, the complaint reads, not present things as they were; distortion will be all but inevitable. In reply – apart from the fact that historians of science have long remained silent about the distortions wrought upon ancient science by translating its contents into the terminology of modern science and mathematics – it should be said that such a complaint criticizes the use of the philosophy of science in precisely that area in which it is likely to prove of greatest value to the history of science. Let us give, for the moment, the devil his due and say that the kind of critical examination that the analysis-synthesis distinction is submitted to in the following pages does not present things *exactly as they were* in all historical detail and nuances. What results is, however, still of great potential value to the historian of science as historian. It seems undeniable that questions, ideas, and factors of the *historical* situation will be brought to light that would have gone unnoticed without the analysis afforded by the present book. And the chances are that these questions and ideas will contribute every bit as much as many other 'non-philosophically influenced' interpretations in approaching the 'definitive' history of the material in question. If this is true – and I believe that it is – then Jaakko Hintikka and Unto Remes have given us a very valuable example indeed of how philosophy can be used to *do* well-informed history of science.

Harvard University   JOHN E. MURDOCH

INTRODUCTION

A reader of this work needs a guide both to the main argument attempted in it and to certain arguments not presented in it.

The main argument is calculated to establish an overall interpretation of the ancient Greek method of mathematical (mostly geometrical) analysis. This method is described briefly in the beginning of Chapter I below. The extensive general description Pappus devotes to it (Hultsch 634–636) is quoted *in extenso* in Chapter II and discussed in detail in subsequent chapters.

Our diagnosis of this method is outlined mainly in Chapters IV–V, and it seeks to present, not just a summary of ancient geometers' theory and practice of analysis, but also and mainly an analysis of the objective sources of the heuristic usefulness of this method. The main theses to which this analysis gives rise are five. (They are of course closely inter-related.)

(i) Logically speaking, a proof obtained by means of the analytical method amounts essentially to a proof by the so-called natural deduction methods (of modern logic). Intuitively, this means that it preferably dealt with a definite configuration of geometrical objects.

(ii) Thus, what was anatomized in an analysis was not a proof (deductive connection), but a configuration ('figure').

(iii) However, this figure cannot always be the one which represents the desired theorem. Typically, auxiliary constructions are also needed. They introduce new geometrical objects into this figure. They are needed because the desired proof or construction cannot be carried out without their mediation. In principle, the main non-trivial, unpredictable element of the analytical method lies in these auxiliary constructions. They are therefore the heuristically crucial but at the same time heuristically recalcitrant element of the methodological situation.

(iv) A straightforward application of natural-deduction techniques to a geometrical theorem, conceived of as an instantiated version of a general implication, would have been to draw conclusions from the antecedent (together with earlier theorems) and at the same time look for the premises from which the conclusion could be inferred. Such a double procedure, whose two prongs can also interact in different ways, was implicitly felt to be too cumbersome, and in practice analysis therefore consisted in drawing conclusions from the conjunction of the antecedent *and* the consequent (both instantiated). It is only this conventional feature that necessitated a separate synthesis which was presented to complement an analysis by justifying the convertibility of its several steps.

(v) Since these steps were of two kinds, deductive inferences and auxiliary constructions, two different types of justification were needed and were in fact presented in synthesis. Moreover, the relative order of these different kinds of steps often had to be changed, which involved further difficulties.

Of these five theses, the connection between the ancient method of analysis and the natural-deduction techniques of modern logic which is mentioned in (i) will be examined systematically in a separate paper, 'Ancient Geometrical Analysis and Modern Logic'. There it will also be shown how an implicit preference of natural-deduction methods is connected with the idea of analysis as an analysis of configurations (cf. thesis (ii)).

The historical aspects of thesis (ii) are discussed in Chapter IV, and further evidence for this interpretation is marshalled in the subsequent chapters, especially in Chapter VIII.

In recent discussions of the ancient method of analysis, a great deal of attention is devoted to the question of the *direction of analysis* as compared with the direction of relations of logical consequence. Does analysis consist of a series of conclusions or does one proceed in an analysis from a hoped-for conclusion to its more and more distant premises? As far as Pappus' general theoretical discussion of the method of analysis and synthesis (Hultsch 634–636) is concerned, what we believe to be a definitive answer is given in Chapter II. There it is shown that Pappus' description presents analysis as an 'upward' movement, that is, as a search of premisses rather than as a sequence of conclusions.

This result at once poses serious further problems, however, especially as it is in conflict with Pappus' own analytical practice. Thesis (iv) is put forward as the main underlying systematical reason for this discrepancy. An explanation is in fact badly needed, for otherwise the presence of synthesis in the consistently two-barrelled method of analysis *and synthesis* – for that is how it is virtually invariably presented and practiced by ancient mathematicians – would remain unaccounted for.

Thesis (iv) is argued for mainly in Chapter IV, where reasons for thesis (v) are also presented. The latter thesis serves to explain the role of the second subpart of the four-part analytical-and-synthetical argument, the so-called 'resolution'. This otherwise very puzzling part of the analytical procedure is examined in greater detail in Chapter VI.

The deepest, and at the same time most elusive, aspect of the method of analysis is undoubtedly the need for auxiliary constructions and their unpredictability. Already the difference between different kinds of argument steps, on which thesis (v) turns, presupposed the need for auxiliary constructions. The underlying reasons for the inevitability of auxiliary constructions, and the methodological consequences of their elusiveness, are mentioned in Chapter I and discussed in the paper, 'Ancient Geometrical Analysis and Modern Logic'. As thesis (iii) brings out (cf. Chapter V), it is precisely the need of auxiliary constructions that imposes the most important heuristic limitations on the very method of analysis, conceived of as analysis of configurations. An interesting corollary to these limitations as applied to the first methodologists of early modern science is mentioned in the last chapter.

We do not pretend that these theses do not leave many loose ends hanging. However, in many cases the very inconclusiveness of the historical material becomes understandable, not to say predictable, on the basis of our analysis of the systematic aspects of the analytical method.

The different Chapters II–IX serve to reinforce these main theses in different ways, at the same time as they contain observations on certain closely related subjects.

Since the only extensive surviving ancient description of analysis and synthesis is due to Pappus, much of what is said in this work takes the form of a series of interpretations of what he says, largely by means of his own practice of analysis, or by relating his ideas to those of other ancient mathematicians and philosophers.

The problem of the direction of analysis presented by Pappus' famous description of the method was already mentioned.

In Chapter III an explicit example of Pappian theoretical analyses is presented and examined, and in Chapters II and VI–VII Pappus' terminology is studied and compared with that of other ancient mathematicians'.

Chapter VIII is dedicated to a study of the ancient tradition of geometrical analysis. It is shown that the ancient ideas about geometrical analysis (including Pappus' account of it) were influenced by different methodological and philosophical views – and by the limitations of these views.

In interpreting Pappus, interesting clues are offered by the terminology of Pappus (and of the other Greek mathematicians). For instance, our conclusions concerning the direction of analysis are largely based on a study of the word ἀκόλουθον (and its cognates) in Pappus and in the texts of other Greek mathematicians. Furthermore, the interpretation of geometrical analysis as an analysis of figures is strongly supported by an examination of the terms ancient mathematicians used to describe its several ingredients (its starting-points, its end-points, etc.). These terms can be seen to refer typically to geometrical objects or other mathematical entities rather than to propositions.

The general philosophical rationale for spilling this much ink on the subject of Greek geometrical analysis is twofold. As explained in Chapter I, it is one of the rare avenues of non-trivial modes of logical reasoning to the theoretical awareness of philosophers of mathematics. It has also served as a paradigm of several crucially important conceptualizations in the history of philosophy and of the methodology of science. Two of these deserve a comment here.

Aristotle's use of the method of analysis as a paradigm of human deliberation is mentioned below in Chapters I and VIII. The later uses of this analogy in the course of history led to developments which are not mentioned below but which enhance its general philosophical significance. Hence a brief explanation might be in order here.

At earlier times, a distinction was often made between knowledge derived from its object and knowledge which determines its object, for instance in the way in which a shoemaker's knowledge of a shoe determines his activity in producing one. The history of this interesting distinction cannot be examined here. It is not difficult to guess, however, that the analytical method, in which one starts from the desired end and aim and tries to reason 'backwards' to ways and means of bringing it about, is as clear and conspicuous an example of such 'intentional knowledge' as one can find in the actual historical material. It is too large a subject however, to be taken up here.

Another use of the method of analysis is briefly studied in Chapter IX. It is the use of the method of analysis as a model of the experimental procedure of the first great modern scientists, notably by Newton. There is a major *lacuna* in our argument here, too. Everybody will admit that some of Newton's major methodological pronouncements made use of the analytical paradigm. From this it does not yet follow, however, that these pronouncements really constituted an adequate representation of his own scientific practice and were not merely *post hoc* rationalizations which maybe only served to guard Newton against certain philosophical objections. We do believe that in a very deep sense Newton really practiced what he preached, and that his methodological pronouncements present an interesting general model of the experimental method at large. We have come to realize that both these claims, also the historical one, need further argument and further evidence before we are prepared to rest our case. Pending further documentation, the reader is invited to consider the last chapter mainly as an object lesson. It serves to show how logical and philosophical analyses of a methodological problem situation can help to adjudicate important methodological claims. In the case at hand, the need and unpredictability of auxiliary constructions in analysis shows once and for all that in spite of its heuristic merits the method of analysis just cannot serve as a foolproof discovery procedure. We hope that this example is sufficient to make also historians of science interested in, if not yet convinced of, the importance of the analytical paradigm for the methodology of early modern science.

The bulk of this book, comprising most of the present Chapters I–VII and IX, was written for the First International Conference in the History and Philosophy of Science in Jyväskylä, Finland, on 28 June–5 July, 1973. (Chapters IV–VI have been considerably revised and expanded, however.) A summary of this material was presented there, and was commented on by Imre Lakatos and Árpád Szabó. A written comment by Szabó is printed below (Appendix I), together with our reply to it (Appendix II). The remaining chapter (Chapter VIII) was written during the second half of 1973.

In a truly co-operative work, it is sometimes hard to tell apart the contributions of its two authors. It may nevertheless be said that Chapters II–VII represent independent work by the junior author, based on the suggestions by the senior author, who is mostly responsible for Chapters I and IX. The background survey presented in Chapter VIII is written by the junior author.

JAAKKO HINTIKKA
UNTO REMES

CHAPTER I

THE HISTORICAL SIGNIFICANCE OF THE IDEA OF GEOMETRICAL ANALYSIS

Analysis (ἀνάλυσις) is a method Greek geometers used in looking for proofs of theorems (theoretical analysis) and for constructions to solve problems (problematical analysis). In both cases, analysis apparently consists in assuming what was being sought for, in inquiring where it comes from, and in proceeding further till one reaches something already known.[1] Analysis is followed by a synthesis in which the desired theorem or construction is established step by step in the usual manner by retracing the stages of the analysis in the reverse order.

This method, which at first blush might seem to belong only to the history of mathematics, has served as a conceptual model for some of the most important ideas in the history of philosophy, including the history of the methodology and philosophy of science. Aristotle compared the structure of human deliberation to that of analysis.[2] Later, geometrical analysis was not only one of the starting-points of Descartes' analytical geometry, whose very name betrays its parentage. It was also one of the inspirations of his general methodological ideas.[3] Newton – probably the most famous admirer of ancient Greek geometry – assimilated in so many words his experimental method to 'the Method of Analysis'.[4] Some historians have seen in geometrical analysis the methodological paradigm of the whole heroic early period of modern physical science from Galileo to Newton.[5] Others claim to have found anticipations of the methods of experimental science, couched in the terminology of analysis (*resolutio*) and synthesis (*compositio*) in the middle of the Middle Ages.[6] Be this as it may, the influence of the analytical techniques of Greek mathematicians on the early development of algebra is unmistakable.[7] Later, the Greek ideas of analysis and synthesis played a role in the formation of Kant's distinction between analytical and synthetical judgements.[8]

This multifarious role of geometrical analysis as a methodological paradigm already fully motivates a close examination of selected aspects of its history. However, for a philosopher of science and of logic there is an additional systematic reason for having a close look at the history of the concept of geometrical analysis. This concept is one of the rare avenues through which instances of non-trivial modes of logical inference have found their way into the awareness of earlier philosophers of science, logic, and mathematics. It therefore is one of the most important links between the systematic interests of those contemporary logicians and philosophers who emphasize the subtlety and power of logical and mathematical reasoning and the traditional philosophies of science and of mathematics.[9]

This point requires an explanation. In fact, it is best appreciated on the basis of some actual work in proving theorems of elementary geometry. Whoever has some experience of this sort, knows that the success of analysis is subject to a certain condition. Speaking first in the intuitive terms referring to geometrical figures, an analysis can succeed only if besides assuming the truth of the desired theorem we have carried out a sufficient number of auxiliary constructions in the figure in terms of which the proof is to be carried out. In principle these may be carried out in the course of the analysis, too, but they can always be accomplished before it. This indispensability of constructions in analysis is a reflection of the fact that in elementary geometry an auxiliary construction, a *kataskeue* (κατασκευή), which goes beyond the *ekthesis* (ἔκθεσις) or the 'setting-out' of the theorem in terms of a figure,[10] must often be assumed to have been carried out before a theorem can be proved. Hence a proof cannot be found by means of analysis without such auxiliary constructions, either.[11]

Several traditional philosophers have keenly appreciated this role of constructions in geometrical proofs. The very terminology of Greek geometers, partly to be discussed later, shows the importance of constructions for proofs in mathematical practice: theorems were called *diagrammata* and proving was called *graphein*.[12] There is a passage in Aristotle which seems to say that once the right constructions have been carried out, the theorem in question is obvious.[13] Later, Leibniz was complaining that "in ordinary geometry, we still have no method of determining the *best constructions* when the problems are a little complex".[14] Elsewhere, Hintikka has shown what role certain generalizations of the geometrical idea of construction played in Kant's doctrine of the synthetical character of mathematical truth.[15]

The topical background of this heuristic significance of constructions is interesting. Most proofs in elementary geometry are essentially first-order (quantificational) arguments.[16] Now already in first-order logic a theorem can often be proved from axioms only by considering more individuals in their relation to each other than are considered either in any of the axioms or in the theorem to be proved.[17] (This can be avoided only by certain artificial tricks, such as coding deductions into what ought to be completely trivial notational features of one's propositions. Such unnatural tricks can be ruled out, and are here assumed to have been ruled out.) What this 'considering so many individuals in their relation to each other' means is easy to understand, although its precise formulation depends somewhat on the particular formulation of first-order logic which is being used. In typical cases, the number of individuals considered together equals the number of individual constants (including the dummy names used in some proof techniques) plus the number of free individual variables plus the number of layers of quantifiers in the sentence in question at its deepest (i.e., the length of the longest sequence of nested quantifiers in it).

What was said earlier is that often this number has to be larger in the intermediate stages of a proof of a theorem from certain axioms than it is either in the theorem or in the axioms. These larger numbers cannot always be predicted recursively on the basis of the theorem, either.[18] In such cases the number of auxiliary individuals needed grows (as a function of the Gödel number of the theorem) faster than any recursive function. In some cases, it can be so predicted, but not in all of them. For theories for which it can be recursively predicted, and for them only, do we have a decision method.

The number of additional 'auxiliary' individuals considered together in a proof may be taken to be a kind of index of the non-triviality of the logical reasoning involved in the proof. (We may perhaps call this kind of non-triviality *combinatorial* to distinguish it from psychological surprise and heuristic fruitfulness.) When no such increase is needed, traditional geometers sometimes spoke of 'logical arguments' as distiguished from geometrical arguments or of merely 'corollarial reasoning'.[19]

These locutions were made possible by a connection between the number of individuals considered together in a formal proof and the number of geometrical objects depicted in the figure by means of which the proof was illustrated. Apart from minor qualifications, the two numbers are essentially the same. An auxiliary construction of a geometrical object by means of certain postulates amounts in an abstract formal proof to drawing an inference (from the existential assumptions that the postulates in effect are) which increases the number of individuals considered together in the proof by one.[20]

Thus the need of auxiliary constructions in geometrical proofs is a reflection of the need of increasing the number of individuals considered together in the proof. It is thus an instance of the combinatorial non-triviality of arguments in first-order logic which was just mentioned. It so happens that in elementary geometry this number is recursively predictable.[21] However, this fact was unknown to all traditional philosophers of mathematics and of science, and is atypical of the general logical situation. If elementary geometry were a 'typical', i.e., undecidable theory expressed in the terminology of first-order logic, this number would be unpredictable by recursive means.

Now all that has been said of the need of considering more individuals together in a *proof* of a theorem from axioms applies *a fortiori* to successful *analyses* calculated to uncover such proofs. In particular, the unpredictability of the number of these auxiliary individuals implies a corresponding uncertainty in the method of analysis. When the unpredictability obtains, one can never be sure in principle that one had carried out enough auxiliary constructions until one has in fact reached the desired result. No *a priori* guarantee of success can be provided. Thus the nontriviality of most logical truths has as one of its consequences an important limitation on the analytical procedure as a method of discovery: it cannot in general be a foolproof (mechanical) one.

This need of auxiliary entities is more conspicuous in the analytical method than in geometrical proofs in general because the very idea of this method was (we shall argue) to consider the interrelations and interdependencies of entities in a definite configuration. (Cf. Ch. IV below, entitled 'Analysis as Analysis of Figures'.) The need of auxiliary entities meant that the relevant configuration usually could not be the one which is given together with the enunciation of the theorem itself (in its instantiated form), but could only be obtained by amplifying this configuration (figure) by means of suitable 'auxiliary constructions'. Hence one of the focal points in this study will be the role of auxiliary constructions in analysis.

Thus the heuristic uncertainty which was noted by earlier geometers and which Leibniz deplored obtains its deep significance from its being suggestive both of the need of constructions, i.e., introductions of new individuals into first-order arguments, and also of the general deductive uncertainty (lack of decision method) which was just seen to be characteristic of even as elementary-looking parts of logic as first-order logic. The need of auxiliary constructions in analysis, and the difficulty of predicting them, is one of the very few situations in which traditional philosophers of mathematics were confronted by the consequences of the non-triviality of logical reasoning. At the same time, the inevitable uncertainty (of sorts) which prevails in the general logical situation of which geometrical analysis is an example is highly relevant to the evaluation of the methodological ideas which in the course of history have been put forward as generalizations of the geometrical method of analysis.

The need of auxiliary individuals can be partly hidden, but not eliminated, by an algebraic treatment of geometrical problems. Instead of carrying out auxiliary constructions, we now have to consider more and more individuals as the values of nested functions with given individuals as arguments, which amounts to precisely the same sort of introduction of new individuals as the use of auxiliary constructions. It may nevertheless be understandable historically that the use of algebraic techniques should have directed mathematicians' attention away from constructions in the geometrical sense of the word, especially in the seventeenth century. However, this only meant that an essential aspect of the conceptual situation was forgotten by them.

These observations do not only motivate an interest in the history of the concept of geometrical analysis (together with its generalizations). They also give it a firmer direction by focusing our attention to those aspects of this history which are most directly related to the systematical problems just adumbrated. It also seems to us that these aspects have not been studied as carefully in the literature as they deserve.

NOTES

[1] Fuller explanations will be given later (see Chapter II).
[2] *Nicomachean Ethics*, III, 3, 1112b15–24.
[3] See, e.g., *Discours de Méthode*, É. Gilson (ed.), Vrin, Paris, 1939, pp. 17, 20.
[4] See Isaac Newton, *Optics*, second edition, Query 23/31, and Henry Guerlac, 'Newton and the Method of Analysis' (forthcoming). Cf. Chapter IX below.
[5] Ernst Cassirer, 'Galileo's Platonism', in M. F. Ashley Montague (ed.), *Studies and*

Essays in the History of Science and Learning, Henry Schuman, New York, 1944, pp. 277–297; J. H. Randall, *The School of Padua and the Emergence of Modern Science* (Saggi e Testi), Padua, 1961, and Oskar Becker, *Grösse und Grenze der mathematischen Denkweise*, Freiburg and Munich, 1961, pp. 20–25.

[6] Cf. A. C. Crombie, *Medieval and Early Modern Science*, New York, 1959, Vol. II, pp. 11–17, 135–146; A. C. Crombie, *Robert Grosseteste and the Origins of Experimental Science*, Clarendon Press, Oxford, 1953, pp. 27–29, 52–90, 193–194, and 297–318.

[7] Jacob Klein, *Greek Mathematical Thought and Origin of Algebra*, The MIT Press, Cambridge, Mass., 1968, especially pp. 154–158; Michael S. Mahoney, *The Mathematical Career of Pierre de Fermat 1601–65*, Princeton University Press, Princeton, 1973.

[8] See Jaakko Hintikka, *Logic, Language-Games, and Information*, Clarendon Press, Oxford, 1973, Ch. IX, 'Kant and the Tradition of Analysis'.

[9] For non-triviality of logic, see Hintikka, *op. cit.*, Chapters VIII and X.

[10] The traditional parts of a Euclidian proposition were (a) enunciation (πρότασις), (b) setting-out or *ekthesis* (ἔκθεσις), (c) account of what is possible or 'specification' (διορισμός), (d) construction (*kataskeue*, κατασκευή), (e) proof (*apodeixis*, ἀπόδειξις), and (f) conclusion (συμπέρασμα). Cf. Thomas L. Heath, *The Thirteen Books of Euclid's Elements*, Cambridge University Press, Cambridge, 1926, Vol. I, pp. 129–131.

[11] Certain qualifications – or perhaps clarifications – are needed here, however. See below Chapter V, p. 46ff. on the sense in which auxiliary constructions are *not* presupposed in analysis.

[12] See here Eckhard Niebel, *Untersuchungen über die Bedeutung der geometrischen Konstruktion in der Antike* (Kantstudien, Ergänzungshefte, Vol. 76), Kölner Universitäts-Verlag, Cologne, 1959.

[13] *Metaphysics*, IX, 9, 1051a23–30. Cf. also T. L. Heath's comments on this passage in *Mathematics in Aristotle*, Clarendon Press, Oxford, 1949.

[14] *Nouveaux Essais*, Book IV, Ch. 3, §6, and Ch. 17, §3.

[15] Hintikka, *op. cit.*, especially Chapters VIII–X.

[16] Cf. Alfred Tarski, 'What Is Elementary Geometry?', in L. Henkin, P. Suppes, A. Tarski (eds.), *The Axiomatic Method*, North-Holland, Amsterdam, 1959, pp. 16–29.

[17] For this paragraph, see Hintikka, *op. cit.*, Ch. VIII.

[18] Otherwise we easily would have decision method for first-order logic, which is known to be impossible.

[19] The former distinction is still used by Heath, *op. cit.* in note 10. For 'corollarial reasoning', see Ch. C. Peirce, *Collected Papers*, ed. by A. W. Burks *et al.*, Harvard University Press, Cambridge, Mass., 1958, Vol. VII, p. 124.

[20] Thus the notion of construction can be made independent of the use of figures to illustrate geometrical proofs.

[21] See Alfred Tarski, *A Decision Method for Elementary Algebra and Geometry*, 2nd ed., University of California Press, Berkeley and Los Angeles, 1951.

It is to be noted, however, that the sense of 'elementary geometry' in which it is decidable is a fairly narrow one, excluding among other things quantification over natural numbers and therefore excluding for instance all theorems about arbitrary polygons. It hence excludes a number of questions in which the Greeks were already interested. But this qualification only strengthens our hand, for we are emphasizing precisely the difficulty of geometrical decision problems.

CHAPTER II

PAPPUS ON THE DIRECTION OF
ANALYSIS AND SYNTHESIS

In spite of the importance of the method of analysis and synthesis for Greek geometry, ancient mathematicians are remarkably reticent about its nature. In fact, several mathematicians of the seventeenth century shared Descartes' belief that ancient mathematicians had on purpose hidden this vital method of theirs.[1] Such paranoic views notwithstanding, Greek mathematicians' quest of formally correct proofs, in conjunction with the general theoretical problems associated with the notion of analysis, amply explains the relative lack of explicit discussions of the method of analysis among ancient mathematicians and philosophers. Analysis is after all a method of discovery, not one of proof.

Although the method was known already to Aristotle – some sources ascribe its discovery to Plato – the only extensive explanation of the concepts of analysis and synthesis is due to as late a writer as Pappus. Fortunately Pappus can be considered a reliable witness. Besides being a competent mathematician in general, he was an accomplished practitioner of the analytical method in particular and also had a thorough knowledge of the history of the method, as his *Collectio* shows.

The status of earlier evidence, especially of evidence dating from the fourth century B.C., is less obvious. The length of the time span between Aristotle and Pappus – and richness of the mathematical activity that had taken place meanwhile – creates a methodological problem. In Pappus, the interesting material is especially relevant because analysis and synthesis are explained by him in connection with the so-called 'Treasury of Analysis' (ἀναλυόμενος τόπος), a collection of doctrines ascribed to mathematicians later than Aristotle. An evaluation of the earlier evidence will only be attempted later. It is in any case reassuring to see that the analyses found, e.g., in Archimedes' works (at least in the form they have been conveyed to us) are essentially similar to the ones carried out by Pappus.[2]

In view of the importance of Pappus' discussion of analysis and synthesis for our appreciation of the nature of these notions, it is in order to quote it in full. We follow the text of Hultsch and use our own translation (which is partly based on those of Ivor Thomas and Thomas Heath).[3]

Ὁ καλούμενος ἀναλυόμενος, Ἑρμόδωρε τέκνον, κατὰ σύλληψιν ἰδία τίς ἐστιν ὕλη παρεσκευασμένη μετὰ τὴν τῶν κοινῶν στοιχείων ποίησιν τοῖς βουλομένοις

634, 5 ἀναλαμβάνειν ἐν γραμμαῖς δύναμιν εὑρετικὴν τῶν προτεινομένων αὐτοῖς προβλημάτων, καὶ εἰς τοῦτο μόνον χρησίμη καθεστῶσα. γέγραπται δὲ ὑπὸ τριῶν ἀνδρῶν, Εὐκλείδου τε τοῦ στοιχειωτοῦ καὶ Ἀπολλωνίου τοῦ Περγαίου καὶ Ἀρισταίου τοῦ πρεσβυτέρου, κατὰ ἀνάλυσιν καὶ σύνθεσιν

634, 10 ἔχουσα τὴν ἔφοδον. ἀνάλυσις τοίνυν ἐστὶν ὁδὸς ἀπὸ τοῦ ζητουμένου ὡς ὁμολογουμένου διὰ τῶν ἐξῆς ἀκολούθων ἐπί τι ὁμολογούμενον συνθέσει· ἐν μὲν γὰρ τῇ ἀναλύσει τὸ ζητούμενον ὡς γεγονὸς ὑποθέμενοι τὸ ἐξ οὗ τοῦτο συμβαίνει σκοπούμεθα καὶ πάλιν ἐκείνου τὸ προηγούμενον, ἕως ἂν οὕτως

634, 15 ἀναποδίζοντες καταντήσωμεν εἴς τι τῶν ἤδη γνωριζομένων ἢ τάξιν ἀρχῆς ἐχόντων· καὶ τὴν τοιαύτην ἔφοδον ἀνάλυσιν καλοῦμεν, οἷον

The so-called Treasury of Analysis, my dear Hermodorus, is, in short, a special body of doctrines furnished for the use of those who, after going through the usual elements, wish to obtain the power of solving theoretical problems, which are set to them, and for this purpose only is it useful. It is the work of three men, Euclid the author of the *Elements*, Apollonius of Perga, and Aristaeus the Elder, and proceeds by the method of analysis and synthesis. Now analysis is the way from what is sought – as if it were admitted – through its concomitants [the usual translation reads: consequences] in order to something admitted in synthesis. For in analysis we suppose that which is sought to be already done, and we inquire from what it results, and again what is the antecedent of the latter, until we on our backward way light upon something already known and being first in order. And we call such a method analysis, as being a solution backwards. In synthesis, on the other hand, we

ἀνάπαλιν λύσιν. ἐν δὲ τῇ
συνθέσει ἐξ ὑποστροφῆς τὸ
ἐν τῇ ἀναλύσει καταληφθὲν
ὕστατον ὑποστησάμενοι
γεγονὸς ἤδη, καὶ ἑπόμενα τὰ
634, 20 ἐκεῖ προηγούμενα κατὰ φύσιν
τάξαντες καὶ ἀλλήλοις
ἐπισυνθέντες, εἰς τέλος
ἀφικνούμεθα τῆς τοῦ
ζητουμένου κατασκευῆς· καὶ
τοῦτο καλοῦμεν σύνθεσιν.
διττὸν δ' ἐστὶν ἀναλύσεως
γένος, τὸ μὲν ζητητικὸν
τἀληθοῦς, ὃ καλεῖται
θεωρητικόν, τὸ δὲ
634, 25 ποριστικὸν τοῦ προταθέντος,
ὃ καλεῖται προβληματικόν.
ἐπὶ μὲν οὖν τοῦ θεωρητικοῦ
γένους τὸ ζητούμενον ὡς ὂν
636, 1 ὑποθέμενοι καὶ ὡς ἀληθές,
εἶτα διὰ τῶν ἑξῆς
ἀκολούθων ὡς ἀληθῶν καὶ ὡς
ἔστιν καθ' ὑπόθεσιν
προελθόντες ἐπί τι
ὁμολογούμενον, ἐὰν μὲν
ἀληθὲς ᾖ ἐκεῖνο τὸ
ὁμολογούμενον, ἀληθὲς
ἔσται καὶ τὸ ζητούμενον,
καὶ ἡ ἀπόδειξις
636, 5 ἀντίστροφος τῇ ἀναλύσει,
ἐὰν δὲ ψεύδει ὁμολογουμένῳ
ἐντύχωμεν, ψεῦδος ἔσται
καὶ τὸ ζητούμενον. ἐπὶ δὲ
τοῦ προβληματικοῦ γένους
τὸ προταθὲν ὡς γνωσθὲν
ὑποθέμενοι, εἶτα διὰ τῶν
ἑξῆς ἀκολούθων ὡς ἀληθῶν

suppose that which was reached
last in analysis to be already
done, and arranging in their
natural order as consequents
the former antecedents and
linking them one with another,
we in the end arrive at the
construction of the thing
sought. And this we call
synthesis. Now analysis is of
two kinds. One seeks the truth,
being called theoretical. The
other serves to carry out what
was desired to do, and
this is called problematical.
In the theoretical kind we
suppose the thing sought as
being and as being true, and
then we pass through its
concomitants [consequences] in
order, as though they were true
and existent by hypothesis, to
something admitted; then, if
that which is admitted be true,
the thing sought is true, too, and
the proof will be the reverse of
analysis. But if we come upon
something false to admit,
the thing sought will be false,
too. In the problematical kind
we suppose the desired thing to
be known, and then we pass
through its concomitants
[consequences] in order, as
though they were true, up to
something admitted. If the
thing admitted is possible or

προελθόντες ἐπί τι
ὁμολογούμενον, ἐὰν μὲν τὸ
636, 10 ὁμολογούμενον δυνατὸν ᾖ καὶ
ποριστόν, ὃ καλοῦσιν οἱ ἀπὸ
τῶν μαθημάτων δοθέν,
δυνατὸν ἔσται καὶ τὸ
προταθέν, καὶ πάλιν ἡ
ἀπόδειξις ἀντίστροφος τῇ
ἀναλύσει, ἐὰν δὲ ἀδυνάτῳ
ὁμολογουμένῳ ἐντύχωμεν,
ἀδύνατον ἔσται καὶ τὸ
πρόβλημα.
τοσαῦτα μὲν οὖν περὶ
ἀναλύσεως καὶ συνθέσεως.
τῶν δὲ προειρημένων τοῦ
ἀναλυομένου βιβλίων ἡ τάξις
ἐστὶν τοιαύτη. Εὐκλείδου
δεδομένων βιβλίον α΄,
Ἀπολλωνίου λόγου ἀποτομῆς
β΄, χωρίου ἀποτομῆς β΄,
636, 20 διωρισμένης τομῆς δύο,
ἐπαφῶν δύο, Εὐκλείδου
πορισμάτων τρία,
Ἀπολλωνίου νεύσεων δύο,
τοῦ αὐτοῦ τόπων ἐπιπέδων
δύο, κωνικῶν η΄, Ἀρισταίου
τόπων στερεῶν πέντε,
Εὐκλείδου τόπων τῶν πρὸς
ἐπιφανείᾳ δύο,
Ἐρατοσθένους περὶ
636, 25 μεσοτήτων δύο. γίνεται
βιβλία λγ΄.

can be done, that is, if it is what the mathematicians call given, the desired thing will also be possible. The proof will again be the reverse of analysis. But if we come upon something impossible to admit, the problem will also be impossible.

So much of analysis and synthesis. This is the order of the books in the Treasury of Analysis: Euclid's *Data*, one book, Apollonios' *Cutting-off of a Ratio*, two books, *Cutting-off of an Area*, two books, *Determinate Section*, two books, *Contacts*, two books, Euclid's *Porisms*, three books, Apollonio's *Vergings*, two books, his *Plane Loci*, two books, *Conics*, eight books, Aristaeus' *Solid Loci*, five books, Euclid's *Surface Loci*, two books, Eratosthenes' *On Means*, two books. In all there are thirty-three books....

The two lines (Hultsch 636, 15–16) which refer to the *diorismos* (see Chapters V and VI below) are omitted here. They seem to be an interpolation.

Pappus' remarks pose several different problems. One of the most obvious ones concerns the direction of analysis as compared with the direction of relations of logical consequence. In recent studies, this problem has usually been formulated as follows. Does analysis consist in drawing logical conclusions from the desired theorem, or in looking for the premisses from which such conclusions (ultimately leading to the theorem) can be drawn? This problem has in fact received a lion's share of attention in recent discussion as compared with the other logical and philosophical problems to which Pappus' discussion gives rise.[4] Mahoney (*op. cit.*, note 4 above; see p. 326) seems to go as far as to assume in effect that this 'directional' problem pretty much exhausts what he calls 'the logical problem of analysis'.

Yet this aspect of the concepts of analysis and synthesis is one of the more superficial ones. In general, preoccupation with the direction of analysis is often a sign that the subtler ingredients of the method of analysis, including the role of constructions indicated above in Chapter I of this study, are being overlooked. In the Middle Ages and in much of the later literature, analysis is simply identified with proceeding 'upstream', in a direction contrary to that of logical or causal implications, while synthesis is identified with proceeding 'downstream' along a sequence of logical inferences or causal connections. This 'directional' sense of analysis and synthesis is but a pale reflection of the richness of the ideas involved in the original Greek concepts, however. This same preoccupation has even led some commentators to misleading interpretations of the Greek method. The consequences of this mistaken emphasis are also seen in the history of mathematics. In fact, the ideas of analysis and synthesis did not become fruitful again after the Middle Ages until the other aspects of the method of analysis and synthesis began to be appreciated, no doubt partly because of the new conceptual sophistication due to increasing mathematical practice.

Yet there are reasons to pay serious attention to the directional problem in Pappus. One reason is that the supporters of the directional view have overlooked certain basic factors of the very directional problem. It seems to us that their difficulties can be solved completely only when other aspects of analysis have become clear. (Cf. Ch. IV below.)

An even more important reason is that Pappus' account seems to be self-contradictory as far as the direction of analysis is concerned. If what happens in the analysis is a transition from the desired result to its

consequences (as τὰ ἀκόλουθα are usually translated), how can we subsequently invert the process and still obtain a series of valid conclusions,
as Pappus' description of synthesis suggests? If *P* logically implies *Q*,
it is not usually true that *Q* logically implies *P*; yet this is what Pappus'
terminology of τὰ ἀκόλουθα understood as 'consequences' seems to
presuppose.

It is true that Pappus' description of an analysis contains a passage
which suggests that it is not for him a transition from the desired conclusion to its logical consequences. This is the account given by him in the
'for'-clause (Hultsch 634, 13–18). It seems to indicate that for Pappus
analysis consisted in assuming the desired conclusion and in examining
how this conclusion might come about, in other words, from what premisses it might perhaps be deducible. But if this account is accepted, then
it becomes a problem how Pappus could describe the very same process
as a transition from the desired result to its consequences (ἀκόλουθα).

Let us call, taking our cue from Gulley, these two apparently incompatible ideas of analysis 'analysis as a downward movement' (i.e., as a
deduction of consequences from the desired conclusion) and 'analysis as
an upward movement' (i.e., as a movement from the desired result to
premisses from which it could be deduced). The problem is how the two
can co-exist in Pappus' description of an analysis.

They could easily co-exist there if it could be assumed that Pappus
thought of all the steps of analysis, conceived of as steps of deduction, as
being convertible (i.e., as being equivalences rather than implications).
This appears to be the only possible refuge to several twentieth-century
writers of the subject, for they have frequently adopted the idea of analysis
as a downward movement. It is in fact true that the general idea of analysis in Antiquity was that of a method of discovering a proof of a desired
conclusion or construction of a desired result; and a chain of steps of
deduction which starts from the desired result obviously can give us this
only if all these steps are convertible.

However, Norman Gulley (*op. cit.*, note 4 above) has presented a most
convincing case against this current interpretation. According to the
external evidence he presents, the prevalent idea both in writers earlier
than Pappus and in later ones was that of analysis as an upward movement.[5] There are therefore very good reasons for thinking that this idea of
analysis as an 'upward movement' is at the very least part of what Pap-

pus had in mind. This is also suggested by the 'for'-clause in Pappus.

Nevertheless Gulley's interpretation suffers from the fact that he cannot reconcile the different parts of Pappus' description. The price he is paying for bringing Pappus into the line with external evidence is to blame him for inconsistency. His thesis is that "Pappus, though apparently presenting a single method with a single set of rules, is really repeating two different accounts of geometrical analysis, corresponding to two different forms of the method...." This conclusion is hard to swallow even for one who has been impressed by the evidence Gulley marshals. For there is no internal evidence for the presence of a confusion in Pappus and the external evidence Gulley gives us would *prima facie* favor the attribution to him of only one of the two ideas, *viz.* that of analysis as upward movement. Surely a case against the current interpretation would be more convincing if we could acquit him of the charge of inconsistency.

Grounds for acquittal can in fact be found. The main stumbling-blocks here are the statements in which Pappus seems to say in so many words that analysis is a transition from the desired conclusion to its consequences. They have been taken by most recent commentators at their face value, which has created the problem of reconciling them with the equally explicit statements which present analysis as an upward movement. We want to suggest, however, that there is no real problem of compatibility here, for the crucial statements are not intended in the sense they have recently been taken. There is no problem here because the statements which seem to present the idea of analysis as a downward movement are intended to be compatible with the statement which present it as an upward path.

This is shown by Pappus' terminology. The terms he uses for what has generally been taken to be the relation of logical consequence and for its *relatum* in those statement which seem to favor the idea of analysis as downward movement are ἀκολουθεῖν and τὸ ἀκόλουθον. It has been argued by Hintikka that in certain passages in Aristotle these words do not refer to logical consequence at all but to something else, and that the precise sense of these words in Aristotle is generally rather fluid and can be gathered from the context.[6] It seems that a case can be made to the effect that they do not refer to logical consequence in the relevant passages of Pappus, either.

In order to see this, the first thing to note is that Pappus' usage is perfectly consistent. Both in the passage where he gives a general description of the method (*vide* Hultsch, 634, line 12) and in the subsequent distinction between theoretical and problematical analysis (636, lines 2 and 9) the words ἀκολουθεῖν and τὸ ἀκόλουθον are *always* used when Pappus is speaking of the path from the desired conclusion to the (unknown) premisses, whereas they are *never* used when Pappus describes the converse path from the premiss to the conclusion we hope to establish. Instead, Pappus uses such terms as ἀπόδειξις (636, lines 5 and 12), ἑπόμενα (634, line 20), and συμβαίνειν (634, line 14).

This dual terminology already suggests a way of reconciling the different parts of Pappus' account. We want to suggest that τὸ ἀκόλουθον in Pappus' description of analysis and synthesis does not mean a logical consequence, but is a much more vague term for whatever 'corresponds to', or better, 'goes together with' the desired conclusion in the premisses from which it can be deduced, perhaps in the sense of enabling one to deduce the conclusion from them. Hence our translation 'concomitant' instead of the usual 'consequence'.

In spite of the difference between our translation and the traditional one, there is in reality nothing extraordinary about our suggestion here. The etymology of the term ἀκόλουθος is said by Plato in *Cratylus* 405 C ff. to be ἀκέλευθος. (This etymology is accepted by Liddell and Scott.) Now Plato interprets the latter term as ὁμοκέλευθος, 'that which (or: he who) has the same way', 'that which (he who) goes together with', exactly as our translation presupposes. The vagueness of this basic meaning of the word squares well with our observations concerning the later history of the meaning of the term (see below).

There is an interesting passage elsewhere in Pappus which strongly supports our interpretation. In Hultsch 30, lines 9–11 we read (the translation is by Ivor Thomas):

ὁ μὲν οὖν τὸ θεώρημα προτείνων συνιδὼν ὁντινοῦν τρόπον τὸ ἀκόλουθον τούτῳ ἀξιοῖ ζητεῖν καὶ οὐκ ἂν ἄλλως ὑγιῶς προτείνοι....	Therefore he who propounds a theorem, no matter how he has become aware of it, must set for investigation the conclusion inherent in the premisses, and in no other way would he correctly propound the theorem....

(Here Ivor Thomas' translation departs a little from the literal one, which would be in terms of seeking τὸ ἀκόλουθον of 'this'. Now the word τούτῳ is an emendation by Hultsch for τοῦτον; the punctuation of the sentence is not unambiguous. If we accept Hultsch' emendation, Pappus will indeed be speaking of the ἀκόλουθον of the theorem.)

Here the crucial word cannot by any stretch of imagination mean a consequence of the theorem to be established (in the axioms). For one thing, many respectable theorems do not imply any axiom or any distinguishable part of an axiom. What is more important, finding an axiom implied by the theorem would not in itself help to establish the theorem. Yet Pappus says that "in no other way would he correctly propound the theorem".

An interpretation of the passage that makes good sense is to understand τὸ ἀκόλουθον to be the 'concomitant' or 'what goes together with' the theorem in the axioms, in the sense of justifying it. This is neither an unnatural nor (for an ancient mathematician) an unfamiliar way of looking at mathematical proofs. Aristotle already discussed on more than one occasion what the 'cause' of this or that geometrical figure's having certain geometrical properties is, finding them in the premisses of the argument that established these properties. "If it [the conclusion concerning the triangle] were not so, then the straight line would not be [as stated in the premisses]" (*Physics*, II, 9, 200a18–19; cf. *An. Post.* II, 10, 94a28–35).

More evidence can be gleaned from Pappus' usage. Hultsch does not seem to list all the passages in which Pappus uses the terms ἀκολουθεῖν and ἀκόλουθον. Those that he does list already yield strong evidence for our interpretation, however. Often Pappus uses these terms quite loosely to indicate various different kinds of 'going together' (other than logical consequence). Cases in point are found in Hultsch 264, line 7, and 252, line 10. In contrast, when Pappus is speaking specifically of a deductive connection (or of a lack thereof), he seems to use other terminology. Thus we find οὐχ ἕπεται for the lack of a deductive connection in Hultsch 254, line 20. There does not seem to be a single passage where a logical consequence is clearly indicated by the word ἀκολουθεῖν in Pappus. The following are examples of other uses.

Hultsch 84, line 7, ἀκολουθεῖν is used in speaking of the following of an example.

At Hultsch 90, line 4, ἀκόλουθος is apparently employed to express 'following suit' rather than logical consequence. At Hultsch 34, 14–15, Pappus again seems to be concerned with a point's following another in a construction.

In a couple of interesting passages (Hultsch 352, line 3, and 864, line 24) ἀκόλουθον or ἀκολουθία is used to express a theorem's being provable in *the same way* as another. There is no indication for (and a rather strong external counter-evidence against) assuming that the first theorem is thought of by Pappus as being a *corollary* of the second. Hence we are again dealing with something different from logical consequence.

These examples suggest that when a term as vague as ἀκολουθεῖν occurs alone, its meaning is likely to vary. It may be that the meaning of this term gets disentangled from deductive following in Pappus largely through the contrast between it and such 'deductive' terms as ἕπεσθαι. If so, we have here essentially the same situation as in Aristotle's *De Interpretatione* 12–13.[7]

Further evidence is perhaps found in Pappus' description of theoretical analysis. At Hultsch 636, lines 2–3, Pappus says that the 'concomitants' (alleged 'consequences') of the thing sought which has been assumed to be true must *also* be assumed to be true and existent by hypothesis (whatever that means). This would be unnecessary if the 'concomitants' were deductive consequences of the thing sought which has already been assumed in the hypothesis. (That deductive inference preserves truth was acknowledged already by Aristotle.)

Once the true meaning of the ἀκολουθεῖν terminology in Pappus is recognised, the apparent inconsistencies in his account which have greatly worried Gulley and others are overcome to a large extent. For instance, the γάρ-clause at 634, lines 13–17, which is (together with lines 17–23) deleted by Mahoney (although it is well supported by the MSS), does not cause any perplexity any longer, for the passage which it introduces is naturally taken to elaborate the brief initial account of analysis Pappus gives at 634, lines 11–13. (Notice that the plausibility of Mahoney's excision is in any case somewhat reduced by the fact that he has to postulate another interpolation at 636, line 3, the words 'by hypothesis'.)

As a piece of circumstantial evidence, it may be noted that ἀκολουθεῖν seems to have been used widely in late ancient mathematics and logic in functions other than that of indicating logical consequence.[8]

Many questions remain, however, even after the main reason for blaming Pappus' account for inconsistency is removed. There is, among other things, a part in Pappus' description which *prima facie* seems to contradict our interpretation of ἀκολουθεῖν. This consists of the very last lines of Pappus' description of theoretical and problematical analysis, respectively. "But if we come upon something false to admit, the thing sought will be false, too" (636, lines 6–7) and "But if we come upon something impossible to admit, the problem will also be impossible" (636, lines 13–14). Here Pappus seems to assume that the truth – or, in the problematical case, the possibility – of the 'admitted thing' would follow from the truth – or possibility – of the thing sought. This assumption would presuppose either the interpretation of analysis as a downward movement, which was already ruled out, or else the reversibility of analysis (cf. Table 1, rows 7 and 8, in Ch. III below).

Furthermore, the dual character of the method Pappus described is a stumbling-block to a simple interpretation. What is involved in his description is a method of analysis *and synthesis*, not a method of analysis alone. This is true both of Pappus' account of analysis and synthesis and of the practice of all known ancient mathematicians. In fact, when analysis and synthesis were later thought of as two separate methods, this is often a sign that the analogy with the Greek geometrical analysis and synthesis is being forgotten or at least somewhat loosened. Examples of this are found in the medieval literature and in the early modern period.

But if we could take Pappus' general characterization of analysis literally (as we have interpreted it), no synthesis would be needed to complement it. For once we have found a premiss from which the desired conclusion follows, and connected this conclusion through a sequence of such premisses with axioms and earlier theorems, no further justification can be required and nothing further remains to be done. Hence our interpretation of Pappian analysis as an upward movement, even though strongly supported by a close reading of the evidence, just cannot be the whole story.

What is even more important, Pappus' general account of analysis and synthesis just is not a faithful reproduction of what he does in his own mathematical practice nor of what other ancient mathematicians were doing. Hence the problem remains of reconciling Pappus' general description of analysis with the actual practice of ancient mathematicians. This will be one of our tasks in the sequel.

It is clear already at this stage of our discussion, however, that there is a little more to be said in favor of the reversibility idea than we have given it credit for so far. Although Pappus' usage shows that ἀκολουθεῖν certainly was not for him a technical term for equivalence, the symmetry of the relation it expresses shows that its use by Pappus is compatible with the idea that the steps of analysis should be reversible. Later (in Ch. IV below), we shall see that there is some further support for this idea in Pappus' geometrical practice, and even in the basic conceptual situation in geometry. It seems to us that Pappus' 'official' description of analysis and synthesis which we have been discussing reflects more his insight into the overall logical situation, where analysis basically means proceeding 'upwards', rather than geometrical practice, where there are forces operative tending to emphasize the downward movement and the reversibility problem.

We can go even further than this. It was not part and parcel of the analytical method as it was practised by Greek mathematicians that each step of analysis should have been put forward as being convertible right from the beginning. Rather, the convertibility was only hoped for at the stage of the analysis, and had to be proved subsequently in the synthesis. This is again shown by the fact that analysis and synthesis were consistently conceived of by the Greeks as the two halves of one single method, not as two separate methods.

Apart from these few problems, and certain others that will be taken up in the sequel, we seem to have made reasonably good sense of the problem of the direction of analysis in Pappus' general description of analysis.

It is in order here to point out also that the whole problem of the direction of analysis has no bearing on the question of the objective sources of the heuristic value of the analytical method. This point is worth emphasizing, for it seems to be misunderstood almost universally. It is often thought that while analysis as a downward movement (deductive procedure) can be rule-governed, analysis as an upward movement must be a matter of 'intuition' and 'divination', to quote terms actually used by Richard Robinson.[9] Nothing could be further from the truth. The rules which tell us when a putative conclusion C actually follows deductively from the known premiss A or from a finite set of known premisses $A,B,...$, also work the other way around, telling whether a putative premiss A (or a finite set of putative premisses $A, B,...$) actually entails a given conclusion C. Since there is no other objective reason in sight for an asymmetry between the two directions, one can therefore proceed upstream by means of the very same chart which shows one how to proceed downstream. (For instance, no preference can be based on numbers here, for in the same way as a given proposition entails several others, in the same way it can itself be inferred from several others.) Consequently, the direction in which one is moving in analysis has nothing to do with the heuristic usefulness of the analytical method. Mahoney notwithstanding (*op. cit.*, p. 323), looking for antecedents is not an intrinsically more 'aimless' procedure (nor a less aimless one, for that matter) than casting about for suitable consequences. Nor is moving in the one direction rather than in the other more conducive to the finding of lemmata to which the desired result could be reduced or to the finding of qualifications to sharpen or to correct this result. One difference between the two directions is of course that moving in one might be more familiar to mathematicians than moving in the other. But this is not an objective difference. However, there may have been somewhat more objective sources of the difference between the two directions. Some of them will be discussed in Chapter IV below.

NOTES

[1] See, e.g., Descartes, *Regulae*, Rule IV (Adam et Tannery, Vol. X, p. 376ff.; cf. p. 373); *Secundae Responsiones* (Adam et Tannery, Vol. VII, p. 156); Wallis, *Mathesis Universalis* (*Opera*, Vol. I, p. 53; cf. *Algebra, Opera*, Vol. II, p. 3). Cf. G. L. Huxley, 'Two Newtonian Studies', *Harvard Library Bulletin* 13 (1959), 348–361, especially pp. 354–355.

[2] *Pappi Alexandrini Collectionis Quae Supersunt*, ed. by Fr. Hultsch, Weidmann, Berlin, Vols. I–III, 1876–1877; see Vol. II, p. 634ff. Cf. *Euclidis Data* (*Euclidis Opera Omnia*, Vol. VI), ed. by H. Menge, Teubner, Leipzig, 1896, p. 235 and p. 252; T. L. Heath, *The Works of Archimedes*, Cambridge University Press, Cambridge, 1897, pp. 61–85, where a number of analyses are found. For the original text, see *Archimedes, Opera Omnia*, ed. by J. L. Heiberg, Teubner, Leipzig, Vols. I–II–III, 1910–1881–1881; Vol. I, pp. 170–210, and Vol. III, pp. 154–172.

In the commentary of Eutocius (Vol. III of Heiberg, *op. cit.*) there are also analyses attributed to persons other than Archimedes, e.g., on p. 92ff. to Menaechmus and on p. 188ff. to Diocles. – All these analyses show convincingly that Pappus' analytical techniques were not his innovation but were rather borrowed from earlier geometers.

There are analyses also in the works of Heron of Alexandria. See *Opera Omnia*, Vol. III, ed. by H. Schöne, Teubner, Leipzig, 1903, 'Vermessungslehre', Book I, Prop.

10 (p. 28ff.); Prop. 11 (p. 30ff.); Prop. 12 (p. 32ff.); Prop. 14 (p. 36ff.); Prop. 15 (p. 40ff.); Prop. 17 (p. 48ff.), and Book II, Prop. 8 (p. 112ff.); Prop. 9 (p. 116ff.); Prop. 12 (p. 122ff.), and Book III, Prop. 4 (p. 148ff,); Prop. 5 (p. 150ff.); Prop. 6 (p. 152ff.); Prop. 7 (p. 154ff.); Prop. 8 (p. 156ff.); Prop. 12 (p. 164ff.); Prop. 14 (p. 166ff.); Prop. 22 (p. 180ff.).

As for Apollonius, we are told by the standard historians that his *Cutting-off of a Ratio* uses analytical method. We have not seen the translation of E. Halley (Latin, from Arabic, 1706), however. This book belonged to the 'Treasury of Analysis', as we have seen. His *Conics* which is also included by Pappus (see above) uses analytical method only exceptionally. But his analyses are similar to the ones found in Pappus and Archimedes, thus supporting our hypothesis of the origin of Pappus' techniques. See *Apollonii Pergaei Quae Graece Extant*, ed. by J. L. Heiberg, Teubner, Leipzig, Vols. I–II, 1891–93. For analyses, cf. Vol. I, p. 274ff.

3 Ivor Thomas, *Greek Mathematics*, Vols. I–II, William Heinemann (Loeb Classical Library), London, 1939–41; see Vol. II, pp. 597–601. Cf. Heath, *Elements*, Vol. I, pp. 138–139. Our Greek text is from Hultsch, *op. cit.*, p. 634–636.

4 M. S. Mahoney, 'Another Look at Greek Geometrical Analysis', *Archive for History of Exact Sciences* 5 (1968/69), 319–348; R. Robinson, 'Analysis in Greek Geometry', in *Essays in Greek Philosophy*, Clarendon Press, Oxford, 1969, pp. 1–15; H. Cherniss, 'Plato as Mathematician', *The Review of Metaphysics* 4 (1951), 395–425; F. M. Cornford, 'Mathematics and Dialectic in the Republic VI–VII', *Mind* N.S. 41 (1932), N. Gulley, 'Greek Geometrical Analysis', *Phronesis* 33 (1958), 1–14; Thomas L. Heath, *Elements*, Vol. I, pp. 137–142.

5 See Gulley, *op. cit.* An especially clear case in point is found in *Procli Diadochi in Primum Euclidis Librum Commentarii*, ed. by G. Friedlein, Teubner, Leipzig, 1873, p. 69; cf. *Proclus, A Commentary on the First Book of Euclid's Elements*, translated with Introduction and Notes by Glenn R. Morrow, Princeton University Press, Princeton, 1970, p. 57.

Aristotle already knew that in an analysis the conclusions do not always convert – and that the cases in which they do not do so are the non-trivial ones. In *Post. An.* I, 12, 78a6ff. he writes: "If it were impossible to prove a true conclusion from false premisses, analysis would be easy, because conclusion and premisses would necessarily reciprocate."

The idea of analysis as an upward movement is also defended by Cornford in *op. cit.* (note 4 above).

6 Hintikka, *Time and Necessity*, Clarendon Press, Oxford, 1973, Ch. III.

7 Hintikka, *Time and Necessity*, Ch. III.

8 Cf., e.g., Nicomachus of Gerasa, *Introduction to Arithmetic*, transl. by M. L. D'Ooge, The Macmillan Company, New York, 1926, p. 291, on ἀκόλουθος; J. W. Stakelum, *Galen and the Logic of Propositions*, Angelicum, Rome, 1940, pp. 72–79. An exception is according to Benson Mates, *Stoic Logic* (University of California Publications in Philosophy, Vol. 26), University of California Press, Berkeley and Los Angeles, 1953, p. 43, the Stoic use of the word.

Could this perhaps explain why Sextus, who was accustomed to the Stoic terminology, apparently accuses geometers of the inconsistency described above on pp. 12–13 of our study? Cf. *Sextus Empiricus*, ed. by R. G. Bury, William Heinemann (Loeb Classical Library), London, Vol. IV, 1949, p. 250, line 14ff.

The Stoic usage seems to have influenced Proclus. Two of his sources (Posidonius and Geminus) were Stoic. At least one occurrence of the word ἀκόλουθον in Proclus can be attributed to Posidonius the Stoic (Friedlein p. 218, line 4; Morrow p. 171, line 14). On the other hand, Proclus' opinions about analysis stem partly from Porphyry, who, although a Neoplatonist, was strongly influenced by Aristotelian ideas of the nature of science. No wonder, then, if Proclus' description of analysis is a mixture of elements which have little to do with actual mathematical practice. The passage at p. 255, line 8ff. (Friedlein) is instructive: it contains the Stoic term μαχόμενον (line 9) and mentions Porphyry by name.

[9] *Op. cit.*, p. 6 (note 4 above). The same mistake is committed, e.g., by R. S. Bluck in his edition of Plato's *Meno*, Cambridge University Press, Cambridge, 1961, pp. 76-78.

CHAPTER III

WHAT PAPPUS SAYS AND WHAT HE DOES: A COMPARISON AND AN EXAMPLE

The agreement between most of the things Pappus says of analysis as well as certain discrepancies between them can be highlighted by studying his several accounts side by side. This is done in the accompanying Table I to which we shall refer back later.

To provide further material for later use and especially to illustrate Pappus' terminology we also give here an example of a (theoretical) analysis in Pappus. It is from Book IV of the *Collectio* (Prop. 4, Hultsch 186, line 9ff.). All geometrical assumptions and geometrical objects are given in every relevant respect in the original form and order. They are formulated in the italicized parts of the argument. Sentences in ordinary brackets serve to fill certain gaps in Pappus' exposition. Some of them are ours, some can be found in Hultsch (*op. cit.*) or in VerEecke's commentary.[1] Sentences in square brackets are our explanations. The division of the argument into steps is somewhat arbitrary. Its main purpose is simply to facilitate later references to the example. (For an example of a problematical analysis, see Chapter VI below.)

I. ENUNCIATION

I(a)   That which is given (the *dedomena*, τὰ δεδομένα).

Let ABG be a circle with the centre E. Let BG be a diameter, and AD a tangent (that meets the circle at *A*); *it meets the extended BG at D. Join D and F. Describe* (the diameter) *AEH. Join H and F; join H and J. The diameter BG meets HF at K, and HJ at L.* (See Figure 1.)

I(b)   The thing sought (the *zetoumenon*, τὸ ζητούμενον).

That EK=EL.

[Meaning, of course, that one has to show that *EK=EL*.]

TABLE I

	General Statement (634, 11–13)	Explanation (634, 13–18)
(1) The starting-point of analysis.	... analysis is the way from what is sought	For in analysis we suppose that which is sought
(2) The kind of initial assumption needed.	– as if it were admitted –	to be already done
(3) Nature of the process.	through its concomitants in order	and we inquire from what it results, and again what is the antecedent of the latter...
(4) Requirements on the intermediate stages.	. . . .	. . . .
(5) End-points of analysis.	to something admitted	until we ... light upon something
(6) Requirements on them.	in synthesis.	already known or being first in order (τάξιν ἀρχῆς ἐχόντων).
(7) Positive outcome.	. . . .	. . . .
(8) Negative outcome.	. . . .	. . . .

Theoretical analysis (634, 26–636, 7)	Problematical analysis (636, 7–14)
(1) In the theoretical kind we suppose the thing sought	In the problematical kind we suppose the desired thing
(2) as being and as being true	to be known
(3) and then we pass through its concomitants in order	and then we pass through its concomitants in order
(4) as though they were true and existent by hypothesis	as though they were true,
(5) to something admitted;	up to something admitted.
(6)	. . . .
(7) then, if that which is admitted be true, the thing sought is true, too.	If the thing admitted is possible or can be done, that is, if it is what the mathematicians call δοθέν, the desired thing will also be possible...
(8) But if we come upon something false to admit, the thing sought will be false, too.	... but if we come upon something impossible to admit, the problem will also be impossible.

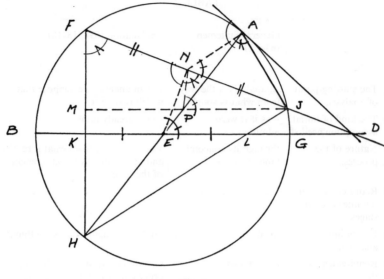

Fig. 1.

II. ANALYSIS IN THE BROADER SENSE

II(a)   Analysis proper (analysis in the traditional sense? The term
          used by Hankel here is 'transformation'; cf. *op. cit.*).[2]

Let it be done (γεγονέτω).

Step₁. *Let a straight line JM, parallel to BG, be described through J*
(it cuts *AH* at *P*). [This is the first auxiliary construction. The *zetoumenon*-
assumption *KE=EL* is not needed.]

Step₂. *MP=PJ* (by construction and because we suppose that
KE=EL). [*Zetoumenon*-assumption and the construction assumed in the
dedomena both used.]

Step₃. *Let a perpendicular from E to FJ be described.* [The *zetoumenon*-assumption not used in this auxiliary construction.]

Step₄. *FN=NJ* (by construction, cf. Euclid's *Elements* 3,3). [The
zetoumenon-assumption not used in this deduction.]

Step₅. (Let *N* and *P* be joined.) [Omitted by Pappus. The *zetoumenon*
not used.]

Step$_6$. *NP and FM are parallel.* (From previous results by *El.* 6,2.) [The fact that $MP=PJ$ is partly based on the *zetoumenon*-assumption and is used here, whereas the fact that $FN=NJ$ which is also used here does not exploit the *zetoumenon.*]

Step$_7$. (Let J and A be joined.) [Omitted; the *zetoumenon* not used.]

Step$_8$. *The angle $JNP=NFM=JAP$* (by *El.* 1,29 and 3,21). [Deduction step as in Step$_6$.]

Step$_9$. *The points J, A, N, P are on a circle* (by the converse of *El.* 3,21). [In spite of its formulation, this step introduces a new geometrical object, *viz.* the circle $JNAP$. The fact that we can introduce this circle is based on the equalities of Step$_8$. If they hold, there is a circle through $JNAP$. Now they were inferred from the parallelism of NP and FM, which in turn was based on the equality $MP=PJ$, as shown by Step$_5$. But this equality was inferred from the fact that the *zetoumenon*-assumption holds, of course together with the original construction and auxiliary constructions based on it. Thus the introduction of the circle $JANP$ is based on the zetoumenon-assumption I(b).]

Step$_{10}$. *The angle $ANJ=APJ=AEL$.* (By *El.* 1,29 and 3,21.)

Step$_{11}$. *A, N, E, D are on a circle* (by the converse of *El.* 3,21). [Why Pappus thinks that they are is not said. However, it is fairly obvious that this last conclusion follows from the equalities of the immediately preceding Step$_{10}$.]

II(b) 'Resolution' (we borrow this term from Hankel).

It is really so (ἔστιν δέ).

R$_1$. *The angles DAE and END are both right angles* (... and the points A, N, E, D therefore lie on a circle). [This observation concerning DAE and END is not inferred from anything established previously in the course of argument, as is shown by two facts. First, if we think that it is inferred from previously gained results, how are we then to interpret the preceding statement 'it is really so'? Secondly, the assumption is totally unproblematic. DAE is a right angle by original construction (cf. I(a)) END is a right angle by an auxiliary construction which depends solely on the *dedomena*. (Cf. Step$_3$.) Clearly a different interpretation of R$_1$ is to be preferred. On it, the fact that A, N, E, D are on a circle is inferred from the *italicized* statement of R$_1$. This statement gives a reason for the fact which is different from the one given earlier in Step$_{11}$. The repetition is made intelligible by the observation that the reason given in R_1 is independent of the *zetoumenon*-assumption – unlike the one given in $Step_{11}$. Hence the angles *DAE* and *END* can be constructed on the basis of the *dedomena* – I(a) – alone; and the proof can therefore begin. – The general conceptual situation is discussed on pp. 43, 54–55 below.]

III. SYNTHESIS

The synthesis is as follows (συντεθήσεται δὴ οὕτως):

III(a)   Construction (κατασκευή).

III(b)   Proof (ἀπόδειξις).

Step[1]. *Because the angles DAE and END are both right angles, A, N, E, D are on a circle.*
[Repeats R_1. Cf. also $Step_{11}$ of analysis.]
Step[2]. *Consequently the angle AND = AED.*
Step[3]. *But because ED and PJ are parallels, the angle AED = APJ.*
[Cf. $Step_{10}$ of analysis.]
Step[4]. *Hence the points A, N, P, J are on a circle.*
[Cf. $Step_9$ of the analysis.]
Step[5]. *Hence the angle JAP = JNP.* [Cf. $Step_8$.]
Step[6]. *But (by the original construction) JAP = JFM.*
Step[7]. *Hence JFM = JNP.*
Step[8]. *Hence FM and NP are parallel.* [Cf. $Step_6$.]
Step[9]. *FN = NJ.* (By *El.* 3,3, for instance.) [Cf. $Step_4$ of analysis.]
Step[10]. *Hence MP = PJ.* [Cf. $Step_2$ of the analysis.]
Step[11]. *And (because) MJ and KL* (are parallel by auxiliary construction), *FH:EH = MP:KE = PJ:EL; on the other hand, MP:PJ = KE:EL; hence KE = EL.*

(Q.E.D.) [ὅπερ ἔδει δεῖξαι, in the MSS usually in the form ὅπερ:∼ which is an abbreviation. Omitted here as in most other cases in Pappus.]

This example prompts several comments. First, it shows what the structure of an analytically established proposition is. The total analytical argument for the proposition we shall call an analytical proof system. It consists of the enunciation of a theorem (or a problem; Part I in our sample analysis), analysis in the broader sense (our Part II), and of synthesis (our Part III). As a rule, an analysis is in Pappus in fact followed by a synthesis.

These main parts of an analytical proof system have in turn their parts. In an enunciation there are two distinguishable parts. First, there is 'the given', which typically catalogues certain geometrical objects and also (usually separately) lists certain assumptions about their interrelations. Then follows the requirement which divides the parts of the enunciation: 'I say that...', 'one has to show that...', 'one has to construct a...', 'one has to find a...', where '...' is to be filled by a description of a geometrical object or of a relation between such objects.

Thus interpreted, Part I is surely for Pappus what Greek geometers called an 'enunciation'. (For instance in our example above it is the only part of the proof system to which this term can naturally be applied.) He speaks of the two parts of such an enunciation as 'the given' and 'the thing sought'. 'The given' here is in reality the result of what is otherwise called 'the setting-out' (*ekthesis*; cf. Hultsch 68, line 19 and 23–24), in spite of the fact that Pappus usually speaks of what is 'given', instead of what is 'set out'. Similarly Pappus usually speaks of our Part I(b) as 'the thing sought', although the same part is called by Proclus the 'specification' (διορισμός, in one of the senses of the word).[3] Thus Pappus' enunciation must be understood in terms of a particular figure.

Occasionally Pappus nevertheless applies the word 'enunciation', like 'the given' and 'the thing sought', also to a substantially different kind of enunciation and to its parts. (This usage coincides with that of Proclus; cf. note 3.) In the case of theorems, such an enunciation is typically a general implication (cf. Hultsch 654, line 25ff.). This kind of enunciation has not yet been 'exhibited' in an *ekthesis*. Its parts (in Pappus, the 'hypothesis', or the 'given', and 'the thing sought') are the 'if'-sentence and the 'then'-clause of the general implication, respectively.

As for problems, the particular ('exhibited') enunciation with two parts is by far the most common in Pappus. The corresponding general enunciation can again be conceived of as a general implication with an antecedent and a consequent. However, this type is very rare in Pappus. What is found more often in Pappus is a simple requirement to construct something (e.g., 'construct a spiral'). It seems that the result of such a required construction was thought of by the ancient mathematicians as 'the thing sought'. Thus this enunciation (of sorts) is closely related with the 'thing sought' of the general enunciation of problems. Although in it 'the given' (and even all references to it) are often missing, for simplicity we shall speak of it as the general enunciation of problems.

Thus there are certain terminological ambiguities here. These ambiguities are not an indication of confusion in Pappus' practice, however. On the contrary, it seems that Pappus' terminology is somewhat oversimplified and lacking in accuracy just because of the practical purpose of smooth exposition.

In an analysis in the broader sense there are again two parts. As far as the problematical analysis is concerned, this fact was recognized long ago. There Part II(a) is often shorter than II(b). This prompted Hankel to think that in the problematical case II(b) is 'the' analysis. He even called it 'resolution', that is, analysis.

In a theoretical analysis there is likewise Part II(b). But now it is the turn of Part II(b) to be insignificant in length. Hence some scholars seem to have assimilated it to II(a) or overlooked it completely (see Chapter II, note 4).

However, both ways of thinking lead us to consider the two kinds of analysis as being structurally different: either the theoretical analysis or the problematical one will lack a part when compared with the other one. This is contradicted by what Pappus says of the two kinds of analysis (see Table I above). It is very difficult to read any structural difference between the two kinds into Pappus' description of them. He clearly thinks of them as having the same structure. In Chapter VI it will be shown that this is the case also in his practice. (Cf. also Part II(b) in our example above and our comments on it.)

In a synthesis of a problem we again find two different parts. (See our example above, III(a) and III(b).) In the few theoretical analyses there are in the *Collectio*, Part III(a), i.e., the construction seems to be missing. This is true of our example, too. (See Prop. 12 of Book IV, Hultsch 204, 4ff.) In the five theoretical analyses found in the scholia to Book XIII of Euclid's *Elements*, the construction is likewise absent.[4] (The same constructions which are introduced in the analysis proper are used in the *apodeixis* of the synthesis.) However, the material is so scanty that caution is in order. Admittedly, the absence of construction in these syntheses *can* have some significance. One might even suspect that the absence of Part III(a) is reflected by Pappus' account on the reverse of analysis at Hultsch 636, 5–6 and 12–13, for he says that the reverse of analysis is the *apodeixis* and does not mention construction at all. However, this argument from silence is demonstrably incorrect in the problematical case (cf. 636, 12–13). (There is almost invariably a construction in the synthesis of a problematical analysis.) Hence the conclusion perhaps should not be drawn in the theoretical case, either. Moreover, Pappus' proofs are admittedly often quite elliptical. In particular, the whole synthesis is missing from some theoretical analyses (Hultsch 906, 26ff., Prop. 157 in Book VII; 908, 22, Prop. 158; 916, 10, Prop. 163). Maybe the construction is missing for the same innocuous reason, that is, because it is so self-evident as to be dispensable.

Tentatively we shall assume that this is so, i.e., that there are two parts in a synthesis of an analytical argument, if complete. In the sequel, we shall also offer a partial explanation of Pappus' statement that the reverse of analysis is *apodeixis* which is compatible with this assumption.

Summing up, we may thus say the following. Usually it is said (see, e.g., Heath, *Elements*, Vol. I, pp. 137–140) that the joint analytical-synthetical procedure has in the theoretical case *two* parts, analysis and synthesis. Now we can see that in Pappus it has four, *viz.* analysis proper, 'resolution', construction, and the proof proper. (He did not use any special terms to refer to the two parts of analysis and to distinguish them from each other, however.) This holds not only of problematical analyses (where the fact was recognized, e.g., already by Hankel and Heath) but also of theoretical analyses. A fuller treatment of the parts of an analytical proof system and of their function in that system will be given in the next three chapters.

NOTES

[1] P. VerEecke, *Pappus d'Alexandrie. La Collection mathématique.* Desclée De Brouwer & Co., Bruges, 1933, *ad loc.*
[2] H. Hankel, *Zur Geschichte der Mathematik in Altertum und Mittelalter*, Georg Olms, Hildesheim, 1965 (reprint of the original edition of 1874), p. 144.
Geometrical analysis and its several parts are discussed also in H. G. Zeuthen, *Geschichte der Mathematik im Altertum und Mittelalter*, Copenhagen, 1874, pp. 137–150. Zeuthen finds Greek names for several parts of the analytical proof system. In our opinion, however, his names are highly conjectural, to say the least, as far as the

parts of analysis are concerned. – It is curious that the complex structure of analytical arguments is overlooked by P. Tannery, 'Du sens des mots analyse et synthèse chez les Grecs et de leur algèbre géométrique', in P. Tannery, *Mémoires Scientifiques*, Vol. III, Paris, 1915, pp. 162–169.

3 Friedlein, *op. cit.* (see Chapter II, note 5), p. 203; Morrow, p. 159ff.

4 *Euclidis Elementa Vol. IV libros XI–XIII continens*, ed. by J. L. Heiberg, Teubner, Leipzig, 1885, pp. 366–376. Heath (*Elements*, Vol. III, p. 442ff.) gives only a sample of the interpolated material. The analyses are, however, to be found in full in many older editions and translations, e.g., in the influential translation of Commandinus (1572).

J. L. Heiberg conjectures (in 'Paralipomena zu Euklid', *Hermes* **38** (1903), 46–74; 161–201; 321–356; see esp. p. 58) that the author of these analyses is Heron. This opinion can be supported by comparing these analyses with those attributed to Heron in an-Nairīzī (*Anaritii in Decem Libros Euclidis Comm.*, ed. by M. Curtze, Teubner, Leipzig, 1899, p. 89ff.). Heron's comments as reported by an-Nairīzī are extremely interesting. He tries to carry out his analyses and syntheses by using as simple a figure as possible (cf. his use of a straight line divided in parts on p. 89). This way of proof is for him a proof 'without the figure' (cf. p. 102, line 3; p. 104, line 3 etc.). This expression is found also in the interpolated analyses in Euclid. (The expression is of course somewhat misleading, because he uses a kind of figure, after all.)

Heron acknowledges, however, that the proof 'without the figure' is not always possible. (Heiberg thinks, falsely, that Heron has the impossibility of *analysis* in mind at an-Nairīzī p. 106, line 11ff. Cf. Heiberg, *op. cit.* above, p. 58.) Cases in point are *Elements*, 2, 1 (a theorem) and 2, 11 (a problem). See an-Nairīzī, p. 89 and p. 106.

In an-Nairīzī the last step of analysis in the broader sense (i.e., including our 'resolution') is apparently repeated as the first step of synthesis (which contains no 'construction'); see, e.g., an-Nairīzī p. 103, 5–8 and 9–19. It is thus clear that our sample analysis of a theorem above is carried out along somewhat the same lines as in Heron, although Pappus does not proceed 'without the figure'. Now the question is: Has this type of analysis other exponents besides Heron and Pappus? We have not found any. Heath, who correctly sees the quasi-algebraic nature of Heron's analytical arguments *apud* an-Nairīzī (*Elements*, Vol. III, p. 442), says elsewhere (Vol. I, p. 378) that Pappus and Heron are the first mathematicians to use semi-algebraic methods. A sound policy, it seems to us, is not to try to seek in these analyses (or in our sample argument above) any general paradigm of the analytical method in the Antiquity. A more representative analytical argument will be reproduced in Ch. VI below. The problematical analysis quoted there, and not our Pappian example above, is of the kind one can find in Archimedes and in Apollonius (cf. Ch. II, note 2). We have nonetheless decided to treat here the kind of analysis our sample analysis above illustrates, not because we think it is representative of analyses in general but because the role of the 'resolution' in it has been overlooked in the literature. Only when one sees that there is a 'resolution' (of sorts, at least) in all analyses in Pappus can one make sense of Pappus' descriptions of theoretical and problematical analysis and understand the symmetry between the two types of analysis in his description.

The passages of Heron *apud* an-Nairīzī cited above can be found also (with substantially the same content as in Curtze *Anaritius*) in *Codex Leidensis* 399, 1, *Euclidis Elementa ex interpretatione al-Hadschdschii cum comm. al-Narizii*, ed. by R. O. Besthorn and J. L. Heiberg (Pars I, Fasc. II, Libraria Gyldendaliana), F. Hegel *et fil.*, *typis excuderunt Nielsen et Lydiche*, Hauniae, 1897.

CHAPTER IV

ANALYSIS AS ANALYSIS OF FIGURES:
THE LOGIC OF THE ANALYTICAL METHOD

Perhaps the most misleading thing about the current discussions of analysis and synthesis in Pappus is that they do not bring out clearly what the analyses of Greek geometers were analyses *of*. The prominence given to the directional problem in fact encourages a misleading interpretation. It tends to suggest that what is being analysed is the deductive leap from axioms to the theorem to be proved, which is analysed into a sequence of steps of deduction (and analogously for constructions in the case of problematical analysis). We shall call this the propositional interpretation or the analysis-of-proofs view. It is our thesis that although it is possible in principle to look upon the method of analysis in these terms, it embodies a wrong emphasis and is consequently badly misleading as to what the actual practice of Greek geometers was and also as to the terminology they employed.

This mistake has probably been encouraged by Aristotle's use of the term 'analysis', as exemplified by his habit of referring to his main logical writings as τὰ ἀναλυτικά, i.e., the *Analytics*. Part of the sense of analysis presupposed here is brought out, as Ross correctly observes, by *Pr. An.* I, 32, 47a2–5. The process in question is the translation of unsystematic verbal arguments into an explicitly syllogistical form (into the moods of the three figures).[1]

This cannot be the whole plot of the two *Analytics*, however, for the *Posterior Analytics* does not deal with the translation of arguments into a syllogistical form. Another thing which is involved there is the resolution of such syllogisms into others. This resolutive process is twofold: a given syllogism may be reduced to a syllogism in another mood or else into a combination of two or several syllogisms. The former kind of reduction is discussed in the syllogistic parts of the *prior Analytics*. In a nutshell, the upshot of the discussion is that syllogistic inference relies essentially on the transitivity of class-inclusion. Hence the resolution of a given syllogism into several (the last of the three kinds of analysis practiced in the two *Analytics*) must proceed by inserting intermediate terms between the major and minor term of the syllogism to be resolved.[2] This kind of resolution of scientific syllogisms into others, and the properties of the ultimate 'atomic' premises which represent minimal syllogistic steps, is Aristotle's preoccupation in much of the *Posterior Analytics*. Here analysis means in a rather vivid sense analysing the deductive step from the minor to the major term by 'bridging' it by means of intermediate terms. This is a variant of what we have called the propositional interpretation of analysis.

The influence of its Aristotelian form was probably enhanced by the fact that Aristotle was well aware of the geometrical notion of analysis (see, e.g., *Eth. Nic.* III, 3, 1112b20ff., *Post. An.* I, 12, 78a6ff.) and that he never makes a clear distinction between it and the other uses of the term 'analysis'.

To the propositional interpretation we want to contrast a different view which may perhaps be called the instantial interpretation or the analysis-of-figures view. On this view, one thinks of the analyst as analysing the figure by reference to which (in case of the theoretical analysis) the proof of the desired theorem is carried out or which (in the case of problematical analysis) exemplifies the several steps leading up to the desired construction.

This point can also be formulated in terms independent of the practice of illustrating geometrical proofs by means of figures. On our view, in theoretical analysis one analyses the complexes of geometrical objects – their interrelations and interdependencies – involved in the proof of the desired theorem, not the deductive steps which would take us from the premises to the theorem; and likewise for problematical analysis. The steps of analysis do not take us from one proposition to another, no matter what the direction of the relation of logical consequence is which obtains between them, but from a geometrical object of a number of geometrical objects to another one.

These formulations of the instantial interpretation are somewhat loose and metaphorical, though closely geared to actual geometrical practice. An important part of our task therefore is to spell out what this interpretation amounts to in sober logical terms.

Some of the difficulties with the propositional interpretation are illustrated already by its relative failure to explain the usefulness of the method of analysis. What is the structure of the proof (synthesis) that

results from a theoretical analysis according to the propositional interpretation? Suppose P_0 is the theorem (proposition) to be proved, and K
the conjunction of suitable (already known) theorems and axioms. The
structure of the deduction which results from the analysis may then be
said to be the following, where $P_1, P_2, ..., P_j$ are suitable intermediate
propositions:

$$(1) \qquad \frac{K, K \supset P_j, P_j \supset P_{j-1}, ..., P_1 \supset P_0}{P_0} .$$

This schema leaves entirely unexplained the role of the desired theorem
P_0 in analysis. Clearly the 'secret' of the method of analysis is to use in
some sense the structure of the theorem to be proved (P_0) to find a proof
for it. Yet there is no room for this in the schema (1). Both K and each of
the links $P_i \supset P_{i-1}$ must be known independently of the analysis. How
are they to be discovered? What role, if any, does the detailed structure of
P_0 play in discovering them? It is not hard to see, it seems to us, that
something has been left out of the picture here.

This was apparently seen already by Hankel.[3] He did not have enough
logical apparatus available to him, however, to be able to spell out the
situation fully. What he tried to do was in effect to represent the theorem
to be proved (our P_0) as an implication (say $A \supset B$), and to represent the
structure of the proof found by means of analysis schematically as follows:

$$(2) \qquad \frac{A \supset E, E \supset D, D \supset C, C \supset B}{A \supset B} .$$

This will not quite do, either. For one thing, very few of the many propositions which can be established through analysis have the explicit form of
an implication. Secondly, it still remains to be explained how the structure
of the desired theorem (the details of what it says of certain entities) helps
us in looking for a proof. Thirdly, if we think of 'what was reached last
[in analysis]' in Pappus' description of the beginning of synthesis (Hultsch
634, line 19) as a theorem (e.g., '$A \supset E$' in (2)) which is known and likewise of the ἑπόμενα (or consequents, that is, '$E \supset D$', '$D \supset C$', $C \supset B$' in
(2)), we have a difficulty. The relation of each of these successive theorems
to its successors in (2) is not one of logical implication, as Hankel already
recognized (cf. *op. cit.* p. 141, line 5: 'reihen' for the linking). Hence the

verb ἕπεσθαι at 634, line 20 cannot stand for deductive consequence on
Hankel's interpretation and on the presupposition that Pappus is speaking
about the steps between theorems. Because the 'deductive' meaning of
ἕπεσθαι in Pappus seems to be well-established (cf. Ch. II), this speaks
against Hankel's interpretation.

Nor is this the last of the problems besetting (2). Admittedly, (2) seems
to capture Pappus' practice somewhat better than (1). But if we accept
this view, we must assume that the ἑπόμενα at Hultsch 634, line 20, and
the προηγούμενα at line 21 are respectively the consequents and antecedents of the implications ('$E \supset D$', '$D \supset C$', and so on), to escape the
objection just registered. Hence the 'consequents' are not theorems, but
antecedents and consequents of theorems which have themselves the form
of an implication.

This leads to another problem, however. On Hankel's interpretation
the analysis connected with (2) must presumably be written (in modern
notation) somewhat as follows:

$$(2)^* \quad \frac{A \supset B, B \supset C, C \supset D, D \supset E}{A \supset E},$$

where '$A \supset B$' is the theorem we are trying to prove, '$B \supset C$',..., '$D \supset E$'
are known theorems (or their substitution-instances) and '$A \supset E$' is (in
the positive case) something which is "in an obvious way true of the
figure A" as Hankel strangely says.[4] If our reinterpretation of (2) is correct and what is linked there according to Pappus are not theorems, we
must reinterpret (2)* in the same vein. This makes nonsense of Pappus'
statement that in synthesis the former 'antecedents' now operate as 'consequents'. For a comparison between (2) and (2)* shows that this is not
the case here. A is here an antecedent only in both (2) and (2)*; B once an
antecedent, once a consequent in (2)*, being always a consequent in (2);
C and D have both functions in both, whereas E, which is a consequent
(but not an antecedent) in (2)*, operates both as an antecedent and as a
consequent in (2). Hence something is again wrong here.

Hankel's views on analysis and synthesis, especially as reinterpreted as
we have done, yield valuable insights into some of our problems. However,
it is clear that they – besides being very crude – rest entirely on the propositional interpretation. Thus we must look for a better interpretation.

THE LOGIC OF THE ANALYTICAL METHOD

In order to see what is going on in Pappus we must realize that a theorem is for him typically a general implication, i.e., is of the form

$$(3) \qquad (x_1) \dots (x_k) \, (A \, (x_1, \dots, x_k) \supset B(x_1, \dots, x_k)),$$

where A and B may still be complex expressions. In case of problems, we are in fact dealing with a proposition of the form

$$(4) \qquad (x_1) \dots (x_k) \, (A \, (x_1, \dots, x_k) \supset$$
$$(\exists y_1) \dots (\exists y_m) \, C(x_1, \dots, x_k, y_1, \dots, y_m)).$$

Now the first step in a Euclidean proposition is an *ekthesis*, the 'exposition' or 'setting out' of the general theorem to be proved. That is, from a statement concerning *any* triangle, *any* circle, or any other geometrical configuration of a certain sort we apparently move over to what looks like a particular case of such a configuration. Usually, this configuration is also assumed to be drawn. It is what the figure depicts by reference to which the rest of the argument is carried out.

In Pappus, the general enunciation is usually omitted, as was already noted, and the theorem itself is given only in an already 'exposed' form.

Logically speaking, all talk of figures and of special cases is of course beside the point. Yet *ekthesis* admits of a characterization in abstract logical terms. In modern logical terms, an *ekthesis* amounts to a step of *instantiation*. What happens is that we move from the general implication (3) to considering the instantiated (free-variable) expressions $A(a_1, \dots, a_k)$ and $B(a_1, \dots, a_k)$. Here $a_1, \dots, a_k$ represents (intuitively speaking) the (indeterminate) geometrical entities which a theorem speaks of (suitable arbitrary lines, circles, etc.) and which are depicted in the figure illustrating the theorem.

What one is trying to find here is thus not so much a proof of (3) from axioms and earlier theorems, but a proof of $B(a_1, \dots, a_k)$ from $A(a_1, \dots, a_k)$. In an analysis, one is essentially trying to bridge this gap by starting from the receiving end, i.e., from $B(a_1, \dots, a_k)$. The important thing is not so much whether one is drawing conclusions from B or trying to find premisses from which (together with A) B could be inferred, but the fact that the logical force of B – what it says of certain kinds of geometrical configurations – is also brought to bear on this task.

In so far as an analysis proceeds 'downwards', its structure can thus be represented as follows, where → represents conclusions initially drawn and ⇢ the conclusions one eventually hopes to establish in the synthesis. Again, K is a conjunction of axioms and suitable earlier theorems. Notice that in the relationships depicted by single arrows the ultimate premiss is the whole conjunction $K\&A\&B$, whereas the double arrows ultimately rely on $K\&A$ as their premiss.

(5)

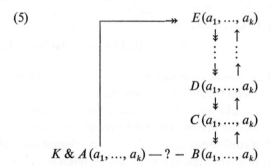

We might thus say that proofs obtained by the ancient analytical method were essentially what a modern logician would call proofs by natural deduction, the hallmark of which is just the kind of systematic instantiation and the use of conditional arguments which have now been found to be characteristic of the analytical method.[5]

It is important to recognize that in practice an analysis in Pappus typically involves drawing the kind of inferences here represented by '→', i.e., inferences from the consequent B which one wants to reach. The idea that analysis can after all proceed downwards thus must be countenanced. The function of a synthesis in the double procedure of analysis-and-synthesis will then be precisely to ascertain that steps downwards can be inverted, i.e., that we can turn each single arrow '→' into a double arrow '⇐'. It is not clear yet, however, why this should have been the uniform practice of Greek geometers, notwithstanding Pappus' admission of the primacy of the idea of analysis as an upward movement.

Even if this admission served merely the purpose of clearing Pappus' conscience as a logician, we have to ask why he departed from the idea of upward movement as a practising mathematician. The crucial point here is that an important aspect of the heuristic usefulness of the method of analysis is due to the possibility of bringing *both* what B says of a certain geometrical configuration *and* what K & A says of it to bear at the same time. This rules out for instance an attempt to proceed consistently upward only, that is, an attempt to try to establish the double arrows directly without using the antecedent 'K & A $(a_1, ..., a_k)$' in the process as a viable heuristic procedure, for then the information of K & A could not be used to see what's what here in order to find the intermediate stages C, $D, ...$ However, what is not ruled out in principle is an attempt to proceed upwards on the right (along the double arrows) while simultaneously applying the force of K & A to the same configuration deductively, i.e., as a premiss. This dual procedure is in principle entirely feasible, and its logic amounts in fact to that of E. W. Beth's method of semantical *tableaux*.[6] However, it is far from surprising that such a two-way traffic was too confusing for practising mathematicians who were impatient with logical subtleties, if not entirely innocent of them. In actual mathematical practice, it was much more convenient to trust one's insight and luck and to proceed downward with full steam (generated both by K & A and by B), damn the torpedoes.

One detail presents a special psychological obstacle to the dual procedure of simultaneously drawing conclusions from A (or from K & A) and looking for premisses from which B could be concluded. It is of course desirable to bring the import of axioms and earlier theorems to bear on the second task and not only on the first. In principle, there is nothing impossible in such an attempt. (One only has to think, e.g., of Beth's method of semantical *tableaux* with its rules for moving formulas, for instance just formulas like our K, from the left-hand column to the right-hand column.) However, psychologically and heuristically this way of using axioms is certainly quite awkward, and was therefore avoided in practice.

Of course this way of thinking made it imperative to justify the procedure *post hoc*. In other words, it made it necessary to have a synthesis to guarantee the reversibility of the several steps. This is not an exorbitantly difficult task, either, for most steps of a geometrical analysis will be reversible anyhow, being mediated by functional dependencies of geometrical entities in a given figure on each other.

This observation helps us to appreciate the significance of the reversibility problem for the practice of analysis. Very likely, part of the inconclusiveness of Pappus' general comments on analysis is due to an unsuccessful effort to reconcile the practical importance of the reversibility problem with the general logical insight that reversibility need not for logical reasons be the crucial aspect of the analytical procedure.

Moreover, it may be noted that synthesis cannot be restricted to the justification of reversibilities. For one of the steps in the synthetic (deductive) argument depicted in (5) (double arrows) is not a reversal of any analytical step, *viz.* the one which takes us from $K \& A$ to B.

All this enables us to see for the first time what is essential and what is inessential in the method of analysis. The most important aspects seem to be (i) the idea of studying the interrelations of geometrical objects in a *given* configuration and (ii) the general heuristic idea of bringing the maximal information to bear on this configuration. The application of both ideas is colored by the fact that geometrical theorems can typically be thought of as general implications. Principle (i) implies, among other things, that we move over to an instantiated version of the enunciation, as indicated. (Hence the importance of the notion of *ekthesis*.) It also implies certain restrictions as to what logical principles are employed in the argument.

Principle (ii) implies that analysis cannot be thought of as a stepwise journey *either* from $K \& A$ to B *or* from B to $K \& A$. Rather, the geometer draws conclusions from K, A, and B together, hoping to turn them later into a proof of B from $K \& A$. But this tentative character of analysis is not essential to it objectively speaking. It is made heuristically natural only by the inconvenience of bringing part of one's information to bear on one's task in one way and part of it in another way. (Precisely the same mild inconvenience easily leads one to prefer in modern logic other methods, for instance Herbrand methods, to Beth's *tableaux* method.)

Thus we reach the important conclusion that from a strictly objective viewpoint neither of the two features which are often thought of as hallmarks of the analytical method touches the gist of its usefulness. These features are (a) its direction (starting from the desired conclusion) and (b) its tentative character which necessitates the use of synthesis as a complement to analysis. Insofar as the former (a) is important, it is important only as a corollary to the principle (ii) of maximal specificity.[7] (Moreover, we saw that in a sense analysis has no direction: it proceeds neither from B to $K \& A$ nor *vice versa*.)

The need of synthesis (cf. (b)) is occasioned merely by considerations of convenience, or perhaps rather a desire to avoid too deep logical subtleties.

Part of what we shall be doing in the sequel is to see how these objective features of the underlying logical and heuristic situation are reflected in the actual historical material.

Notice also that there are structural differences between the analysis of deductive steps and the analysis of figures. The former is usually thought of as linear, and can always be made linear in a natural fashion. The natural development of an analysis of figures is not linear but takes the form of more complicated network of connections. For instance, to determine one of the sides of a triangle, one may have to determine its two other sides and an angle, each through an independent sequence of interconnections of the different elements of a configuration. Here we have, not a linear structure, but no less than three 'lines' of interconnections. These several lines of thought can be pressed into the form of a linear argument, but only by somewhat unnatural means.

NOTES

[1] See W. D. Ross, *Aristotle's Prior and Posterior Analytics*, Clarendon Press, Oxford 1949, pp. 1–2. This use of the term may be borrowed from geometrical usage; cf. B. Einarson, 'On Certain Mathematical Terms in Aristotle's Logic', *American Journal of Philology* 57 (1936), 33–54 and 151–172, especially pp. 36–39. Cf. also following passages of the *Analytics*: 49a18; 91b13; 43a16–24; 46b38–47a9; 52b38–53a3.

[2] See Jaakko Hintikka, 'On the Ingredients of an Aristotelian Science', *Nous* 6 (1972), 55–69.

[3] Hankel, *op. cit.*, p. 139ff., esp. p. 139, line 7.

[4] Hankel, *op. cit.*, p. 140. What is strange in this pronouncement of Hankel's is that *A* of course is not a *figure* at all, but a statement about one. In general, Hankel's account of theoretical analysis is apparently supposed to be based on the geometrical practice of the ancients. It is so brief that it is not easy to judge its adequacy. For instance, it is not clear whether Hankel recognized that there is a second part in the theoretical analysis in the broader sense. (Cf. p. 140, lines 36–37.) His failure to see the similarity between his 'theoretical analysis' and the 'transformation' of 'problematical analysis' in the analytical proof system suggests that he did not.

[5] This connection may be pushed further. The assumption (which seems to be implicit in the traditional literature) that in an analysis all the requisite constructions can be carried out in the beginning of the argument, reflects the possibility of certain normal forms of natural-deduction proofs (or proofs by closely related methods) which are characterized by the very fact that the introduction of new entities is accomplished in the beginning of the argument. For such normal forms, see for instance William Craig, 'Linear Reasoning', *Journal of Symbolic Logic* 22 (1957), 250–268.

More generally, this connection between natural-deduction methods and the traditional concept of analysis seems to make possible comparisons between old discussions of geometrical heuristics and recent studies of the mechanical theorem-proving.

[6] This is not strictly true, for in principle an analyst could try to see what the preceding stages $C(a_1, ..., a_k)$, $D(a_1, ..., a_k)$ are from which $B(a_1, ..., a_k)$ could be deduced while keeping all the time an eye on the fact that K & $A(a_1, ..., a_k)$ is supposed to hold. This is nevertheless awkward for several reasons. A geometer is perhaps somewhat unaccustomed to proceed upwards against the direction of logical implications. What is much more important, it is extremely hard heuristically to do so while keeping in mind the distant goal K & $A(a_1, ..., a_k)$ one wants to reach. The 'information' given to an analyst by K & $A(a_1, ..., a_k)$ can only be put to effective use by drawing conclusions from it (jointly with $B(a_1, ..., a_k)$), hoping all the time that all the steps could be reversed.

[7] The fruitfulness of paying attention to the sought-for conclusions (or, in algebra, to the unknowns) in solving mathematical problems of different kinds is stressed by G. Polya in his well-known works on mathematical heuristics. A brief summary of some of his ideas is given in *How To Solve It* (Princeton University Press, Princeton, 1945). *Respice finem*, he tells us (p. 195). What Polya says of heuristics in mathematics squares (on the whole) well with what analysis (as analysis of figures) does. He is also aware of the similarities between the method of analysis, as described by Pappus, and his own approach (*op. cit.*, pp. 129–136).

CHAPTER V

THE ROLE OF AUXILIARY CONSTRUCTIONS

Our analysis of the logical and heuristic basis of the method of analysis can be pushed further. Perhaps the most important part of it was the *prima facie* dependence of the steps of deduction used in an analysis *both* on *K & A and* on *B*. What is even more striking is that the same tentative dependence of the intermediate steps on the desired conclusion obtains also in the area of auxiliary constructions as was noted earlier in the case of other steps of analysis. These constructions in fact introduce an element into the overall logical situation we have not yet taken into account.

As was already hinted at in the first chapter of this study, often an argument cannot be successfully conducted without auxiliary constructions, that is, without considering more individuals at the intermediate stages between $A(a_1, ..., a_k)$ and $B(a_1, ..., a_k)$ than in either of them. The true picture of the logical situation will then be something like the following

(6)

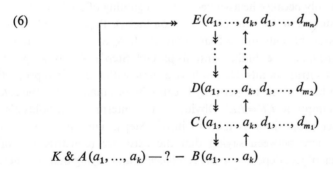

$$E(a_1, ..., a_k, d_1, ..., d_{m_n})$$
$$D(a_1, ..., a_k, d_1, ..., d_{m_2})$$
$$C(a_1, ..., a_k, d_1, ..., d_{m_1})$$
$$K \& A(a_1, ..., a_k) - ? - B(a_1, ..., a_k)$$

There is also no great secret as to how $d_1, d_2, ...$ are introduced. Ancient geometers would have said that this happens by means of postulates. In the eyes of a modern logician, these postulates are of the form

$$(z_1) ... (z_p) (\exists u_1) ... (\exists u_r) (F(z_1, ..., z_p) \supset C(z_1, ..., z_p, u_1, ..., u_r))$$

What happens in a proof is now that we first substitute some free terms we already have at hand for $z_1, ..., z_p$ and then instantiate with respect to $u_1, ..., u_r$, in order to introduce new ones.

The striking thing here is that sometimes the desired conclusion $B(a_1, ..., a_k)$ is used as a guarantee that this process works, i.e., that the substituted terms (we may think of them for simplicity as $a_1, ..., a_k$) satisfy F. We have an instance of essentially just this phenomenon in our sample analysis above in Chapter III. We said there that Step$_9$ introduces a new geometrical object, *viz.*, the circle *JANP*, which depends in a sense explained there on the desired conclusion. Pappus' formulation of Step$_9$ is somewhat elusive, however. It seems to suggest that a *theorem*, not a postulate or a problem as might be expected, is being relied on. Hence this step needs a closer scrutiny.

Let us suppose that Pappus proceeds in the simplest way and in fact infers the italicized conclusion in Step$_9$ by evoking the converse of third proposition of Book III of Euclid's *Elements*. As Heath in effect points out in his commentary (*Elements*, Vol. II, p. 50), one first has to *construct* a circle passing through three of the points *J*, *A*, *N*, *P*. This construction introduces a new entity into the proof, and is always possible to carry out. Then Pappus can prove that the fourth point is on this circle (cf. his formulation in Step$_9$). But he can prove this at the present stage of his argument only because he assumed in the beginning of analysis the *zetoumenon*-assumption (cf. our comments on Step$_9$). Pappus needs a circle which passes through all the four points *J*, *A*, *N*, and *P*, as is shown by the proof he later gives in the synthesis part (cf. Step^4). This is not possible with four arbitrary points. He can have a circle with the desired properties already in his analysis proper only because he originally made the *zetoumenon*-assumption '$EK = EL$'. Obviously, this interpretation holds also if some other proposition is used to justify Step$_9$. Thus there is no real difference here between saying that the existence (construction) of a geometrical object depends on the *zetoumenon* and saying that it is merely this object's having the desired properties that depends on the desired conclusion.

An auxiliary construction depends on the desired consequence also in the analysis part of Proposition 12 of Book IV of *Collectio* (Hultsch 204, 4ff.). In Pappus' problematical analyses the dependence of auxiliary constructions on the sought-for constructions is even more blatant, and much more frequent. Cases in point are for instance the problems at Hultsch

142, 1ff. where it is required to inscribe regular solids in a sphere. First it is assumed that the sphere and all the angle points of the solid in question meet; the auxiliary constructions exploit then this (so far) hypothetical assumption.

This possible dependence of the auxiliary constructions (instantiations by means of postulates) on the desired conclusion B is important for our appreciation of what Pappus in fact does. It means that we need in a combined analytical-synthetical procedure more than the synthesis which shows that the ordinary deductive (non-constructional) steps of analysis can be inverted. These are typically inferences based on the properties of identity (cf. Euclid's axioms or 'common notions') and on propositional logic. Their convertibility is established in the proof (*apodeixis*). (Cf. our example in Chapter III.) But the possibility of carrying out the steps of construction (instantiation) used in the analysis independently of the desired conclusion has also to be established. This is what happens in Hankel's 'resolution', exemplified by II(b) of our sample argument above. Once the possibility is established (by examining, intuitively speaking, what depends on what among the different geometrical objects $a_1, ..., a_k$, $d_1, d_2, ...$) the actual construction (introduction of $d_1, d_2, ...$) is carried out in the *kataskeue* (first part of the synthesis) in such a way as to be based on A and K alone.

Thus the elimination of a possible dependence of *both* constructions (instantiations) *and* other steps on the desired conclusion is what motivates Pappus' four-part structure of analysis (analytical proof system).

This analysis of the logical form of an analytical proof system shows why, and in what sense, it can be described as an analysis of figures. Once all the instantiations are carried out, all inferences we can draw concern the geometrical objects $a_1, ..., a_k, d_1, d_2, ...$, more specifically, what can be said of them on the basis of A, K, and B. What this amounts to in intuitive terms is typically to study how these geometrical objects depend on each other. In modern 'analytical' (algebraic) terms this corresponds to setting up enough equations connecting the unknowns (determined by B or figuring in the auxiliary constructions) with the known elements of the situation (specified by A).

However, a successful analysis usually cannot be an analysis of the 'figure' (geometrical configuration) of the kind mentioned in the premiss A of the general implication that the desired theorem usually is. It is an analysis of this figure amplified by suitable auxiliary constructions. In the light of these facts, a geometer's attitude to auxiliary constructions easily becomes somewhat ambivalent. They represent, for the very good reasons indicated in Chapter I above, the unpredictable element in geometrical analyses. In looking for a proof analytically, one can never be sure that one had carried out enough constructions. In general, the choice of suitable auxiliary constructions is the third and perhaps the most important essential feature of the heuristic situation one encounters in applying the analytical method, a feature quite as important as the two principles (i)–(ii) mentioned earlier on p. 38.

Its significance is perhaps more easily overlooked than that of the other principles (i)–(ii). One small reason for this is the fact that the complication of the figure representing A may be due, not to explicitly mentioned auxiliary constructions, but to conjoining A with suitable earlier theorems K. For the geometrical configurations they speak of may very well involve more geometrical objects than those with which A speaks. This way of introducing new individuals as it were by stealth does not change the methodological situation in principle, however, although it does show that the decision as to which earlier theorems a geometer is to bring in is not always a trivial matter.

The methodological contrast between auxiliary constructions and the rest of the argument is in fact striking. Even though Aristotle was wrong in saying that a proof is obvious as soon as the appropriate constructions have been carried out, it is clear that the rest of a geometrical argument can be systematized in various ways, especially by bringing in algebraic means for discussing the interdependencies between the different geometrical objects. This development is what eventually led to Descartes' 'analytical' geometry. Sometimes the main attention of geometers was in fact focused on the study of these dependencies, and auxiliary constructions were seen as a necessary evil.

These constructions may in any case be thought of as being carried out before the analysis proper begins. The extreme form of this emphasis is what might be called the constructional sense of analysis and synthesis. The distinction between analysis and synthesis is then no longer a difference in direction, but lies rather in the fact that in analysis no constructions are carried out whereas a synthesis virtually amounts to the very carrying out of geometrical constructions. This is roughly the sense in which geometers today speak of analytical and synthetical methods, respectively. The usage is not in itself helpful for the purpose of studying the views of ancient geometers for they knew perfectly well that one cannot usually dispense with auxiliary constructions. Even when they tried to push constructions out of the 'analysis proper' so as to be performed before it, practising Greek mathematicians could not fail to see in them an interesting and non-trivial element in the situation. In fact, it was the early modern mathematicians who tended (in their enthusiasm over the use of algebraic methods in geometry) to forget the constructional aspect of the situation, not Greek geometers.

Some light may nevertheless be thrown on geometers like Pappus from this point of view. At Hultsch 636, 5–6 and 12–13 Pappus does not refer to the reverse of analysis (proper) as synthesis, as might be expected.[1] Instead, Pappus says that "the proof will be the reverse of analysis". Our analysis-of-figures interpretation shows what his point is, however. If analysis is a series of steps which start from those parts of the figure which illustrate the desired theorem, and which establish connections between these and certain – as it were – pre-existing entities, we of course do not obtain a synthesis in the sense of construction by simply reversing the order of these steps. (Cf. Ch. III, p. 28, however.) What we obtain is rather the proof (ἀπόδειξις), in the precise technical sense of that proof proper part of a Euclidean proposition which follows the *ekthesis* and the auxiliary construction (κατασκευή). In such an ἀπόδειξις no constructions are carried out any more than in an analysis in its constructional sense.

Pappus' keen awareness of the role of auxiliary constructions is also shown by his statements at Hultsch 636, lines 1–2, that in theoretical analysis what is being sought is assumed to exist and to be true (ὡς ὄν καὶ ὡς ἀληθές). This two-barrelled statement can be understood as postulating not only the truth of all assumptions and the existence of all individuals mentioned in the enunciation of the theorem to be proved but also the existence of enough of the sought-for auxiliary constructions which would enable us to prove it. This is not belied by the fact that some auxiliary constructions used in analysis occasionally depend on the assumed truth of the desired consequence as noted above and illustrated by Part I(b) of our sample analysis.

It must be realized, of course, that the dependence of the analytical procedure on constructions is twofold. The eventual proof will depend on the actual constructions carried out in the synthesis. This deductive dependence is mirrored by a dependence of the analytical search for a proof (or means for a proof) on suitable auxiliary constructions carried out typically as a part of analysis or preceding it.

Thus from the point of view of the definitive deductive argument auxiliary constructions are presupposed in analysis as it were only hypothetically, not actually. (It may even happen that a tentative auxiliary construction eventually turns out to be impossible!) On the contrary, the very purpose of analysis is to find the desired constructions which is executed in the synthesis. This holds obviously for problematical analysis, where the construction constitutes a solution to the problem at hand. It is also true of theoretical analysis, where the proof cannot be carried out before appropriate auxiliary constructions have actually been carried out in synthesis. Notice, in this connection, how in Pappus' general description of analysis and synthesis, intended to cover both problematical *and* theoretical analysis, it is said at Hultsch 634, line 22, that synthesis finally leads "in the end at the construction (*kataskeue*) of the thing sought". Thus in the most literal sense constructions are not presupposed in analysis, but are rather sought after in it, and carried out in the synthesis. It is a subtle characteristic of the analytical method, however, that it can succeed in finding these *actual* constructions only if enough as it were *hypothetical* ones were already anticipated in it.

Earlier, we saw that the ill-named 'resolution' embodies a clear recognition on Pappus' part of the tentative (hypothetical) character of the auxiliary constructions used in the analysis proper.

We may put the point as follows. Just because in the proof of a theorem auxiliary constructions are often vitally needed, the aim of analysis is the discovery of these auxiliary constructions. However, the actually needed auxiliary constructions can be discovered analytically only if enough such auxiliary constructions have already been hypothetically assumed in the analysis.

Because of the need of auxiliary constructions in theoretical analyses, such analyses are precisely as unpredictable and as difficult to turn into a foolproof method as problematical analyses. Conversely, our examination of the structure of different kinds of analyses earlier in the present chapter shows that problematical analyses can be described in the same logical terms as theoretical ones. Methodologically and epistemologically, the two kinds of analysis are therefore on a par, as they indeed are in Pappus' description. In a theoretical analysis one is looking for a proof of a conjectured theorem. In a problematical analysis, one is looking for a way of carrying out a conjectured construction. Neither is a method of discovering new theorems or constructions more than the other.

Because of this parity, we find Mahoney's assimilation of the difference between theoretical and problematical analysis to a distinction between theorem-proving and problem-solving misleading and even pernicious. There is no truth to his implied allegation (*op. cit.*, p. 320) that *diorismos* could not be treated in formal terms or that, "although theoretical analysis can be described in logical and epistemological terms as a method, problematical analysis cannot" (p. 330). Logically, epistemologically, and heuristically neither kind of analysis is any better or any worse off than the other.

What Mahoney is rightly trying to do is to do justice to the heuristical aspects of the analytical method. However, the way to do so is not to create arbitrary distinctions between theoretical and problematical analysis, but to realize the structural reasons for the fruitfulness of the analytical procedure and also to appreciate the deep connection there is between the nature of the analytical method (especially the role of auxiliary constructions in it) and the nontrivial character of logical and mathematical truth at large.

This point can perhaps be generalized. One of the most interesting general observations one can apparently make here is that the structure of the analytical method, which both nowadays (cf. Mahoney, *passim*) and at earlier times (cf. Descartes and his contemporaries) has often been thought of as a predominantly heuristic method, nevertheless can be analysed in explicit logical terms. This analysis does not preclude the possibility of still viewing geometrical analysis as deriving its value from its heuristic suggestiveness. Instead, we seem to have here an opportunity of understanding in more precise terms than before the very nature of this heuristic process used so successfully in logic and in mathematics. The historical significance of the analytical method of the Greeks adds to the promise of this application of insights gained in logic and foundational studies.

Such applications depend heavily on our recognition of the non-trivial nature of the logical and mathematical truths, however. It is this non-triviality that opens the door to the possibility of locating the precise reason why the analytical method does not, its fruitfulness notwithstanding, yield a mechanical (effective) discovery procedure. This reason lies in the need of auxiliary constructions and, more specifically, in the recursive unpredictability of the number of constructions needed. Instead of overlooking the heuristic and creative element in mathematics (in the case at hand, in geometry), our modern tools enable us to appreciate it even better. As indicated above (Ch. IV, in note 5), it even seems possible to connect discussions of the analytical method known from the history of mathematics with recent studies of the heuristics of formal theorem-proving. Further work is nevertheless needed in both directions.

NOTE

[1] Mahoney finds this an indication of inconsistency which, *inter alia*, prompts him to conjecture an interpolation in the text. All support in the MSS is missing, however. Cf. *op. cit.*, p. 325 (see our Ch. II, note 4).

CHAPTER VI

THE PROBLEM OF THE 'RESOLUTION'

1. THE 'GIVEN' TERMINOLOGY

It was recognized long ago that we can distinguish two parts in a problematical analysis in the broader sense (our 'analysis proper' and 'resolution'), even though no such distinction is explicitly made by Pappus himself. In the present work, we have already seen how the functions of analysis proper and 'resolution' differ and that the latter serves to establish the independence of constructions needed in solving the problem or proving the theorem from the assumption to be proved. However, there remains the multiple task of providing further evidence for this interpretation and of discussing certain residual problems concerning the nature of the 'resolution'. Because of this multiplicity, this chapter naturally falls into several sections. Our discussion will also lead us to comment on certain general features of ancient mathematics and philosophy which form the background of our problem.

A good starting-point is offered by Pappus' use of the word δοθέν ('given'). It helps to relate Pappus' general description of analysis and synthesis with his own analytical practice. Its occurrence in Hultsch 636, line 11, shows that very likely it is also the 'resolution' of problematical analysis that is referred to at 636, 10–14. This word and its cognates are in fact part and parcel of Pappus' terminology in his 'resolutions' of problematical analyses, occurring in them literally hundreds of times. As an example one can use Prop. 33 of Book IV of the *Collectio* (the 'resolution' begins at Hultsch 278, 11). There one can also see the typical use of the propositions of Euclid's book *Data* in the 'resolution' of a problematical analysis. Hultsch lists six applications of the theorems of the *Data* in the 'resolution' of this single problem. This role of the *Data* can probably be connected with the fact that Euclid's book is mentioned first by Pappus among the works belonging to the *Treasury of Analysis* (Hultsch 636, line 19). Marinus in his commentary on the *Data* says: "... the knowledge of the 'given' is simply necessary in the so-called Treasury of Analy-

sis."[1] Also Pappus has written a (lost) commentary on the *Data*. Marinus
says: "The kind of teaching in it [*viz.* in the *Data*] is not synthetical, but
analytical, *as Pappus rightly showed in his commentaries on this book*"
(our italics).[2] These facts, combined with the observation that there seems
to be a 'resolution' in all analyses in Pappus, seem to entitle us to connect
the 'given' mentioned in his general description of analysis (at 636, 11)
with his use of the 'given'[3] in the actual 'resolutions'.

In connection with the term 'given' Pappus refers in so many words to
the usage of mathematicians (Hultsch 636, 11). Hence we must see what
the usual application of the word was. In Pappus and elsewhere in mathematical texts the term has two closely related uses.

In Euclid's *Data* something can be given in three ways: in magnitude,
in species, and in position. But here it is more important to know what
this 'something' can be, in other words, what can be 'given'. In the *Data*,
what is 'given' is always a geometrical or at least a mathematical object,
such as area, line, angle, rectilinear figure, point, circle, segment, magnitude, or ratio.[4] Pappus follows this usage closely, in conformity with the
usage of other mathematicians. Thus what is 'given' is not a theorem or
axiom (as the interpretations of certain modern commentators seem to
presuppose). In typical cases, it is a geometrical object. The use of the
term in the *Data* can be illustrated by a theorem of this book: "If from a
given point a straight line is produced at *given* angles to a *given* straight
line, the straight line in question is *given* in position." (*Data*, Prop. 30,
p. 53, our italics.) Theorems like this is what Pappus appeals to in the
'resolutions' of problematical analyses (see below).

This whole usage is an excellent illustration of what we meant by saying
that for Pappus analysis was primarily an analysis of figures, not of
deductive connections. For what is being analysed in an analysis is of
course mainly what is 'given' in it. And this, we can now see, amounts to
configurations of geometrical objects, not to propositions. And the same
goes for most other uses of the same terminology in Pappus.

Another use of the word 'given' is more specialized. Here, too, the form
δοθέν, or δεδομένον, or other forms of δίδωμι are used. In this sense,
we are told, the things that are given are the same sort of entities as in the
first case. But they must be elements of the configuration which the
antecedent of the enunciation of a theorem or a problem deals with.[5]
This is, according to Marinus, what it means to be 'given by hypothesis'.

Marinus says that this specification of what is 'given by hypothesis' does not apply to the usage of the *Data* because it "is faulty with respect to the consequent (*zetoumenon*)". Marinus prefers an other definition of the 'given by hypothesis': it is "... that which is considered analogically with the *arkhai* [*viz.* the antecedent of an enunciation]".[6] This probably means that what is 'given by hypothesis' can also include the premisses of a resolutive step taken in the midst of an argument (cf. the reference to the 'antecedent' above). In fact, it typically happens in the arguments of the *Data* that as the proof goes on, what formerly was not 'given' (i.e., neither given by hypothesis nor proved to be given before the step in question) becomes part of what is 'given' ('given by hypothesis', according to the definition preferred by Marinus) for the next step of the argument (see, e.g., Prop. 31, *Data* p. 53).

Because the second use of the 'given' terminology is relative to a hypothesis, different things can be taken to be 'given' in different parts of the same argument, provided that the relevant hypothesis is chosen differently. Thus in an analysis proper, both the entities mentioned in *K & A and* those mentioned in *B* could in principle be taken to be 'given', whereas in the 'resolution' and in the synthesis only the configuration specified by *K & A* (or perhaps that specified by *A* alone) was initially taken to be 'given'. Of these two possible uses of the term 'given', Pappus generally follows the latter only.

Thus we can distinguish what is given ('given by hypothesis') in the antecedent of an enunciation from what can be (or is) proved to be given on the basis of what was originally given. These senses, and these two only, can be found in the *Collectio*, too. Other mathematicians used the term δοθέν in the same way (see, e.g., Archimedes' terminology,[7] and Plato's *Meno* 87A).

The word πορiστόν which appears as a synonym of δοθέν in Pappus' description at Hultsch 636, 11 is also found in Marinus' commentary. πορiστόν is "what is not yet done but can be done (πορiσθῆναi)".[8] Furthermore, some people had defined δοθέν as πόρiμον. Thus we have found again a connection with Pappus' description of analysis and the usage of the *Data*.[9] Moreover, πόρiμον is, according to Marinus, "what we can do or *construct* (our italics)". This blunt reference to the construction[10] (κατασκευάσαi, cf. κατασκευή) is, of course, in complete accordance with what we said of the typical application of the word δοθέν to geometrical objects.

A word of warning may be in order here. Although the 'given' terminology is characteristic of Pappian 'resolutions', the same terminology
can of course occur in other contexts, too. The very purpose of studying
the kinds of existential or constructional interdependencies conveniently
codified in this terminology can be different from justifying the auxiliary
constructions used in analysis proper.

2. A SAMPLE 'RESOLUTION'

In order to see the role of a 'resolution' in an analytical proof system, we
translate a problem of Book VII of the *Collectio* (Prop. 155, Hultsch 904,
line 17ff.). The main parts of the argument are enumerated as in our
sample analysis of a theorem (pp. 22–26 above). [11] The translation is
in italics.

    I(a)   That which is given.

*Let a segment of a circle be given, with the chord AB. Let a ratio be given.
Inflect* (cf. Figure 2)

    I(b)   The thing sought

... into the segment two straight lines AC, CB in the given ratio.

    II(a)   Analysis proper.

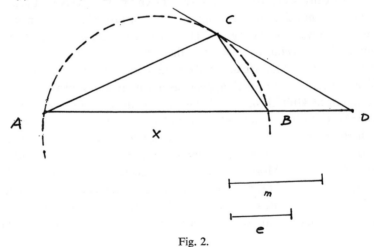

Fig. 2.

Let it be done.
Let a tangent CD from C be drawn;
$AC^2 : CB^2 = AD : DB.$

II(b)   The 'resolution'.

But AC:CB is a given ratio (cf. I(a)); hence $AC^2 : CB^2$ is given; hence the
ratio AD:DB is given. And the points A and B are given; hence the point D
is given, and the tangent DC (Data Prop. 91); hence the point C is given.

III(a)   Construction of the synthesis.

The synthesis is as follows.
Let the segment be ABC, and the ratio e:m. We make $AD:DB = e_2 : m_2$.
We draw through D the tangent DC; I say that the straight lines AC, CB
solve the problem.

III(b)   *Apodeixis* of the synthesis.

Because $e_2 : m_2 = AD : DB$, and $AD : DB = AC_2 : CB_2$ (CD is a tangent by
construction – cf. III(a)), $e_2 : m_2 = AC_2 : CB_2$; *hence $e:m = AC:CB$; thus*
the lines AC, CB solve the problem.

This example vividly illustrates our earlier point that in problematical
analyses the 'resolution' served to establish the independence of whatever
constructions were needed from the desired one, i.e., in Pappus' own
terminology, to establish that all the geometrical entities needed in the
argument were 'given'.

3. 'RESOLUTION' AS A PART OF ANALYSIS

The example likewise shows that the 'resolution' was considered a part of
analysis by Pappus. The synthesis begins only after the 'resolution' (cf.
the typical beginning of synthesis in our example above: "the synthesis
is as follows"). Hence the 'resolution' does not belong to this part of the
argument. In fact sometimes the 'resolution' is referred to by Pappus as
'analysis'.[12]
    The analysis proper is likewise sometimes referred to simply as 'anal-
ysis'.[13] Thus we can find a certain looseness and vacillation in the use of

the word 'analysis' by Pappus. (We may also note that in the expression 'Treasury of Analysis', ἀναλυόμενος, this looseness goes so far that a whole body of books consisting of different kinds of arguments and not of analytical ones only forms the 'Treasury'.) Hence it is correct to speak of II(a) + II(b) together as analysis in the broader sense.

In spite of the fact that the 'resolution' is often longer than the analysis proper, it has an entirely subordinate role in Pappus' general description of analysis and synthesis, in the sense that little explicit attention is paid to it. From Table I we can see that it is not even mentioned when the analysis is described in general terms (above, p. 23). In the accounts of the two different kinds of analysis one may perhaps see an indirect reference to it, for the starting-points of the 'resolution' are of course also the endpoints of the analysis proper, and the word 'given' (and other words for the endpoints of analysis in Pappus' description) can refer to objects having this double function in practice. (See Table I, rows 7–8.)

But the heuristic nature of analysis (attributed to it by Pappus at Hultsch 634, 6 as well as by tradition; cf. Klein *op. cit.* in Ch. I, note 7; p. 261, note 220) apply to analysis proper rather than to the 'resolution'. For in a 'resolution' no new constructions are carried out, but old ones justified, while it is the finding of the right constructions that is heuristically crucial. Similarly Pappus' remarks about the direction of analysis vs. *apodeixis* (as a reverse of it) apply better to the analysis proper. In Pappus' description of analysis, the place of honor is therefore occupied by the analysis proper. How, then, can the 'resolution' be a part of analysis? As far as we can see, no specific Greek names for the several parts of analysis (in the broader sense) are to be found in Pappus. We shall venture an answer to this question only at the very end of this chapter.

4. WHAT 'RESOLUTION' DOES AND DOESN'T ACCOMPLISH

The 'resolution' helps to illustrate the nature of the analysis proper, too. For it is at once clear that the role of the 'resolution' in fact is an auxiliary one when compared with the analysis proper, and that the subject-matter is the same in both in the sense to be explained. If we return to our sample problem above, we see that the analysis proper introduces certain new objects (the tangent CD and points C, D) and makes an assertion concerning the interdependencies between the new objects and those parts of the figure which were originally specified in the enunciation of the problem. It assumes that the problem is possible and in fact is solved, that is, the requirement on the *zetoumenon* (I(b)) is fullfilled. Now the 'resolution' shows that the very objects which were introduced in the analysis proper (being of the same kind and having the same relations as in it) can be constructed on the basis of the information given in the *dedomena* part of the enunciation (I(a)).[14] Thus the 'resolution' shows that the constructions in analysis proper are independent of the *zetoumenon*. It must be stressed that the 'resolution' typically does not yet prove that these constructions actually solve the problem. The interdependencies found in the analysis proper may depend on the *zetoumenon* (cf. p. 42). What is required in a proof is a 'reverse' deduction from the *dedomena* (and known propositions) through the interdependencies to the *zetoumenon* (in the schema (6) on p. 41, from K & A through $E, ..., D, C$ to B following the double arrows. This reversibility cannot be assumed without proof. The ancient geometers admittedly sought to prove the reversibility of the deductive (non-constructional) steps, but only in the *apodeixis* of the synthesis. As for the 'resolution', it does not even attempt to prove this reversibility. On the contrary, it accepts without qualification those properties which the geometrical objects introduced in the analysis proper were there assumed to have. What the 'resolution' proves is that these hypothetical objects having these so far hypothetical properties can be constructed on the basis of the *dedomena* alone. But if the deduction invented in the analysis proper is not reversible, the proof fails, even if the construction is possible on the basis of the *dedomena* alone. (That is, this particular attempted proof fails. This can happen when the construction is irrelevant – or insufficient – for the proof, or when the problem is either partially – in some cases satisfying the data of the problem – or totally impossible.) Of course, in simple cases (e.g., in the sample problematical analysis above) the reversibility may be of little importance in practice. But clearly Hankel is right when insisting that the 'resolution' only serves to justify the requisite *construction*; it does not show that the problem is solved thereby (cf. section 6 below).

In principle, an attempted *apodeixis* of a synthesis can thus fail even after a successful analysis (in the broader sense), it seems. We have found no actual examples of this in the literature, however. Aristotle says, it is true, that we can sometimes analyse the figure although we cannot carry out the synthesis (*Soph. El.* 16, 175a28). This piece of evidence is conclusive, however, only on the presupposition that the analysis of Aristotle's time and the analysis we find in Pappus have the same structure. In itself, this presupposition seems to be consistent with the evidence. (For instance, in both kinds of analysis the subsequent synthesis seems to be needed, and can fail after a successful analysis.) But Pappus' 'Treasury of Analysis' is later than Aristotle; in particular, the 'resolution' might be a later refinement of the analytical technique. It need not be, although the *Data* of Euclid constantly used in Pappus' 'resolutions' is later than Aristotle. But the 'resolution' may have been different in nature in its infancy (perhaps more informal, or used in different ways). At this stage we can neither accept the evidence of Aristotle at its face value nor reject it altogether; its relevance remains to be studied (cf. the beginning of Chapter VIII below).

The absence of unsuccessful syntheses after successful analyses (in the broader sense) in the texts can be explained, however. A mathematician usually addresses only definite results to a wider public. If an analysis was successful but the proof failed, the search for a proof was normally continued until a proof one way or another (i.e., a proof of possibility or a proof of impossibility) was found. In fact, we find references to situations in which the synthesis did not initially succeed but was completed only after the discovery of suitable new propositions (cf. *Apollonius*, ed. Heiberg, Vol. I, p. 4, line 10ff. and Vol. II, p. 178, line 15ff.). In these cases, the analysis was apparently carried out first. (Nobody seems to have been worried about it but only about the synthesis.) Such passages also show that the class of propositions used in actually successful analyses was sometimes not extensive enough for the purpose of completing the synthesis. (This shows that a synthesis could in principle fail even after a successful analysis, as we have suggested.)

5. 'RESOLUTION' AND THE NEGATIVE OUTCOME OF ANALYSIS

We have so far spoken about what a successful 'resolution' can accomplish (together with a preceding analysis proper) in practice. This case is rather subtle. For instance, after successfully completing the whole analytical argument we apparently could easily supplement the 'resolution' by the materials from the *apodeixis* so that it would be a self-supporting argument. This shows that in principle ancient analysts could have dispensed with the synthesis. It nevertheless seems that they were not wont to do so.

But what about a negative outcome of analysis? Can the 'resolution' serve to show that a problem is impossible? A custom connected with the 'resolution' seems to justify this question. The custom is that the questions of solvability (of *diorismos*, see Hultsch 30, 14–16, including partial and – as a limiting case – complete impossibility) were often studied in this part of the total system. An instance is found at Hultsch 48, line 12ff.: "If this is not assumed, the solution is impossible by plane methods; this is clear for everybody... by the numbers in the analysis." However, here Pappus does not prove the impossibility he is studying but only refers to the failure of the efforts of his predecessors (Hultsch 44, lines 18–20). In general, if the 'resolution' proceeds as Pappus' practice suggests, it cannot establish the impossibility of a problem (or the falsity of a theorem). As we have seen, the 'resolution' has a special form. It turns on steps like 'if this is given, that is given' (where what is 'given' is a mathematical object). Even if we have exhausted all possible ways of trying to carry out one of the auxiliary constructions on the basis of the *dedomena* (I(a)) alone, through the mediation of known theorems of the form 'if this is given, then that is given', and failed to obtain a desired justification for the construction, even then we have not proved the impossibility of the problem in question. We have not even proved that we cannot carry out the particular auxiliary construction in question, for the possibility is still open that it could be obtained by means of suitable further additional auxiliary constructions which are not yet being considered. This possibility clearly follows from our analysis of the logical structure of an analytical proof system above. This possibility is also vividly illustrated by the proofs of the theorems of Euclid's *Data*. Many of the proofs given there involve constructions (e.g., Prop. 67, p. 122ff.). Thus the need of new constructions in the 'resolution' was avoided by reference to theorems like those established in the *Data*. But the supply of these theorems being (essentially) limited, the failure in justifying a particular construction by means of all available theorems of this kind was not necessarily final, and could perhaps be remedied only by means of suitable auxiliary constructions. The need of auxiliary constructions in the 'resolution' could be avoided only by moving the problem of additional auxiliary constructions to the proof of an additional theorem belonging to the *Data* (or by somehow taking care of the problem already in the analysis proper).

Thus the 'resolution' alone is incapable of showing the impossibility of a problem. We shall return to the relation of analysis – including analysis proper – and of proofs of impossibility in the next chapter, and discuss there also Pappus' statements about the different possible outcomes of analysis at Hultsch 636, lines 4–7 and 10–14.

6. WHAT 'DIORISMOS' DOES AND DOESN'T ACCOMPLISH

Hankel thinks that while the 'resolution' alone cannot prove the problem to be solved, in conjunction with the διορισμός ('restriction') it can do this: "However, the question [as to when a problem is possible and when it is not] can be solved completely in the analysis."[15] He thinks that the synthesis is redundant in the analytical proof system when there is a restriction in the analysis (and a proof for it). Hankel even is led to ascribe this 'redundancy' to the national fondness of the Greeks of 'unnecessary' logical rigour. However, here Hankel clearly misses the point. Probably he is following Proclus: "... *diorismoi*, whose purpose is to determine when a problem under investigation is capable of solution and when it is not..." (tr. by Morrow, p. 55, lines 5–6). However, the redundancy of synthesis is not entailed by such descriptions of the *diorismos*.[16] Even if we know on what general conditions a problem is solvable or a theorem provable, we do not know yet whether it is solvable by means of those particular constructions and deductive steps which are undertaken in some particular analysis. This still remains to be shown in the synthesis.

In a slightly different vein, it may be noted that in an analytical proof system the *diorismos* follows immediately the 'resolution'. Hence, all that can be established by the *diorismos* is that if the problem is solvable, then certain connections must obtain between the different 'given' elements. However, as long as it has not actually been shown that those particular constructions that are in fact used in the analysis proper are successful, i.e., that all the deductive steps in the analysis can be converted, it has not been shown that the problem is solvable as soon as these conditions on the 'given' are satisfied. And this can only be shown in the synthesis. In brief, a *diorismos* of the kind practised by Pappus can only establish *necessary* but not *sufficient* conditions of solvability in advance of a synthesis. Hankel is therefore mistaken in thinking that the *diorismos* can establish when the problem is solvable. That the *diorismos* in practice produces only necessary conditions seems to have been recognized also by Heath.[17]

We acknowledge, of course, that there is a problem of redundancy in analytical arguments. For as far as merely giving the proof is concerned (and not the way in which it is discovered), the synthesis part can do all that is needed, including the job of the 'resolution'.

7. THE PROBLEM OF THEORETICAL 'RESOLUTIONS'

It is perhaps thought that what has been said of the 'given' and of the 'resolution' applies to problematical analysis only and not to the theoretical analysis. There are in fact several problems of interpretation here. However, the two types of analysis do not seem to us to differ essentially from each other in this respect, notwithstanding the ubiquity of the 'given'-terminology in the 'resolution' parts of problematical analyses (cf. Hultsch 636, line 11, and observe that the 'given' terminology is there used by Pappus only when speaking of the problematical analysis).

The analyses proper of these two kinds are hardly distinguishable. In three important theorems of the *Collectio* (they are Pappus' *lemmata* to the *Surface-loci* of Euclid) the 'resolution' is also essentially like that of a problematical analysis.[18] As in the problematical analyses, there is also a separate construction in the synthesis.

In the commentary by Pappus on the *Almagest* of Ptolemy we likewise read: "... we analyse [this] theorem of the *Almagest* in this way." The analysis seems to be the kind of 'resolution' with the typical use of the 'given' terminology which we are accustomed to expect in problematical analyses.[19]

Thus the features which might seem most characteristic of problematical analyses are found in both main kinds of analyses. Perhaps this is not entirely unexpected. There is admittedly a real difference between theorems and problems (cf. above p. 35). Often this difference was recognized by the ancients, and problems were correctly characterized as questions concerning existence (or concerning processes of construction).[20] This division of geometrical propositions into problems and theorems is acknowledged by Pappus (Hultsch 30). But some of the earlier geometers had emphasized the similarities between these two types of propositions, and Pappus was aware of this.[21]

There seems to have been a general discussion about the nature of geometrical reasoning among the followers of Plato.[22] All participants seem to have admitted that two elements are involved in such reasoning: proof and construction. As for their relative value some (notably Speusippus) emphasized the role of proof and thought that all propositions can be conceived of as theorems.[23] Others, notably the school of Menaechmus, emphasized the constructive element, and were willing to call all propositions problems. As Proclus observes, both parties are right. Concerning the doctrine of the Menaechmians Proclus says: "... the discovery of theorems (*heureseis*) does not occur without recourse to matter."[24]

This amounts to an emphatic recognition of the necessity of using auxiliary constructions in proving theorems. We have already seen that this recognition is firmly grounded on the actual logic of the method of analysis, and that there are in fact auxiliary constructions in theoretical analyses (cf. our example in Chapter III).

8. THEORETICAL 'RESOLUTIONS' OFTEN RUDIMENTARY

This recognition is also mirrored in the 'resolutions' of a number of Pappian analyses of theorems. However, the analyses of theorems described above (that is, those using the 'given' terminology) do not exhaust our diet of examples. The other 'resolutions' give rise to considerable problems of interpretation. It is unlikely that analyses of the sort just described were the only ones in Pappus' mind when he gave his general description of theoretical analysis. For, as we have seen, all references to the 'given' are missing there (cf. Table I, rows 7–8). This is in keeping with the fact that there is a group of theoretical analyses where the 'given' terminology is likewise missing. (Our sample theorem in Ch. III belongs to this group.)

The problem of interpretation is here lent a special poignancy by Pappus' general description of analysis and its varieties. The structural similarity of his descriptions of the two kinds of analysis given there leads one to expect that the same structural similarity between theoretical and problematical analyses can be found in Pappus' practice, too (cf. Table I, p. 23). In the analyses of our last group (for simplicity, from this point on we shall speak in this chapter of the method employed in the analyses of this group as '*the* theoretical analysis') this expectation is only partially fulfilled. The analysis proper, it is true, is here similar to that of all other analyses (e.g., in it there are auxiliary constructions). However, after a theoretical analysis there is no construction part in the synthesis (see above p. 28). Even more conspicuous is the fact that the 'resolution' of a theoretical analysis is different from the 'resolutions' of problems. Not only is the 'given'-terminology not used, but the entire 'resolution' is in theoretical analyses so brief that it easily remains unnoticed.

But even in a theoretical analysis there nevertheless *is* an (at least rudimentary) 'resolution' present. To that extent the expectations aroused by Pappus' general description are therefore justified. Witness, for instance, our example on p. 25 above. The 'resolution' there opens with the formula 'it is really so' (ἔστιν δέ) which is characteristic of theoretical 'resolutions'. That particular 'resolution' begins very much like the 'resolution' of a problematical analysis. It starts from the *dedomena* part of the enunciation (and thus proceeds in a direction opposite to that of analysis proper). It, too, tries to make the argument independent of the *zetoumenon* part of the enunciation. But in this special case the beginning is also the end: there is only one step.

Moreover, even this single step seems to have a role somewhat different from that of the corresponding step in a problematical analysis. The scarcity of material does not justify any categorical conclusions here. It nevertheless looks as if, whereas in the problematical case the 'resolution' is only an attempt to show that the constructions are independent of the *zetoumenon*, here the 'resolution' seems to duplicate the first step of synthesis. There are in fact auxiliary constructions in the preceding analysis proper, and the step in question seems to establish interdependencies between these and other parts of the configuration (cf. p. 24., esp. $Step_{11}$ and R_1) of the kind that might be expected, although the theorems of the form 'if this is given, that is given' are not used. But it seems that if the 'resolution' were to proceed, the result would be an *apodeixis*. A clear case in point is Prop. 12 of Book IV. (The 'resolution' is at Hultsch 206, lines 7–11: the synthesis follows immediately.) Other examples, few as they are, offer some evidence corroborating this impression. In them, too, the 'resolution' of a theoretical analysis begins with the formula 'it is really so' and takes off from the *dedomena* but goes on merely to anticipate the first step of the proof in the synthesis or to indicate that that first step can be carried out.[25]

It is not very surprising that the first step in the synthesis should have been accorded a special position by Pappus. For it is seen from our schema (5) (p. 36 above) that this first step (i.e., the step from K & A to E) is different from the rest, for it does not amount to the converse of any step in the analysis (as we already pointed out). Moreover, it was usually left tacit in the analysis proper. Hence we perhaps should not be entirely surprised that it became part of the task of the 'resolution' to justify this exceptional step of the synthesis half of an analytical proof system. It is readily seen that this was in any case part of the task of the 'resolution', and apparently sometimes almost the whole task.

This interpretation would of course also explain why there is only one step in the 'resolution' of a theoretical analysis: further steps overlapping with the *apodeixis* had not the special role of the first step, and they could be dismissed.

9. 'RESOLUTION' AND APODEIXIS

The rudimentary character of theoretical 'resolutions' admits in many cases a simple but powerful explanation. In theoretical analysis the *zetoumenon* is typically a relationship between several geometrical objects, whereas in a problematical analysis it is a geometrical object (its existence or its constructibility, whichever expression you prefer). Now an auxiliary construction carried out in an analysis proper may depend crucially on the presence of a geometrical object. However, a construction rarely depends entirely on a particular relationship between geometrical objects, in the sense that no constructions of the same kind could not be carried out. For instance, in our sample analysis the only constructions which depended on the *zetoumenon* were the constructions of circles through for points J, A, N, P, and A, N, E, D, respectively. That there were such circles was originally established on the basis of the *zetoumenon*. However, a circle can be drawn through any three of a given collection of four points. (This assumption of existence – or constructibility – is not problematic.) Then what remains to be established is that it passes through the fourth one, too. But since this fact does not any more involve a construction (for the fourth point can be assumed to be given), its vindication (independence of the *zetoumenon*) will be established in the proof in any case. Thus what remains for the 'resolution' is merely the role of a 'discussion-stopper' in analysis, if we want to preserve the parity with the problematical analysis and insist on introducing a 'resolution' in the theoretical case, too.

This observation can be generalized. Often, probably always, there is in theoretical analyses a kind of parity between what can be established in the 'resolution' and what is established in the *apodeixis*: the former can be traded for the latter. This explains why theoretical 'resolutions' got mixed up with syntheses and why their identity was vaguer than that of problematical 'resolutions'.

10. 'RESOLUTION' AND THE ORDER OF AUXILIARY CONSTRUCTIONS VIS-À-VIS DEDUCTIVE STEPS

These observations do not destroy the parallelism of theoretical and problematical 'resolutions'. For instance, the two have in any case the same direction in Pappus' practice. Hence many of the same things can be said of them. (As to the questions of provability, the rudimentary 'resolution' cannot be more effective than the 'resolution' in the problematical case.) However, the problem remains as to why Pappus should have dealt with his theoretical 'resolutions' as casually as he apparently did.

This subordination of the 'resolution' to the synthesis (especially to the *apodeixis*) in theoretical analyses may be partly explainable the result of the incomplete mastery by Greek geometers like Pappus of the delicate logical problems involved in turning analytical arguments into synthetic ones. This pertains especially to the relative order of constructious (existential inferences) and deductive steps in a proof. As a consequence of his intellectual insecurity in this particular domain, Pappus played it close to the chest and let the 'resolution' of a theoretical analysis follow as closely as possible the familiar, safe synthesis. For it is in theoretical analyses rather than in problematical ones that the interplay of deduction and construction is the most conspicuous, and the most puzzling.

This is in fact a subject in connection with which deeper studies in the proof theory of elementary geometry would greatly help an historian of the analytical method. Few such studies exist, however, and none can be attempted here. We can only sketch the outlines of the situation.

As was already noted in Chapter I, an analysis can only be successful
if enough constructions are carried out before it or in the course of it.
One natural possibility is to carry out all requisite constructions before
the rest of the analysis. Although this possibility is made especially natural
by the idea of geometrical analysis as analysis of figures, it was not normally utilized by Greek geometers.

This is all the more remarkable in view of the fact that in a Euclidean
proposition auxiliary constructions were normally separated from the
'proof proper' part and placed in the separate *kataskeue* preceding it.
(From a proof-theoretical viewpoint, this represents a nontrivial normal
form of geometrical proofs.) One reason why no such segregation was
practised in analysis proper is that such a placing of the constructions
would have made the 'resolution' especially difficult. The kind of justification which the 'resolution' aims at can be obtained only if the constructions involved take place first in the order of synthesis, before the deductive steps expounded in the *apodeixis*. (For otherwise some steps of
apodeixis might be needed to establish the feasibility of these constructions, which contradicts the fact that *apodeixis* follows after the 'resolution'.) But this apparently would mean carrying out the auxiliary constructions only at the very end of analysis, not before but rather after the
'analysis of figure' was already accomplished. This is not only unnatural
but usually impossible, if the analysis is to be successful.

The solution of this apparent paradox lies of course in the fact that
logically speaking a 'resolution' often involves more than merely reversing
the order of steps followed in the analysis proper. It involves in effect
changing the relative order of different kinds of steps of argument, in
particular the relative order of steps of construction *vis-à-vis* deductive
steps.

Such changes are often possible, but they are rarely trivial. Every
student of contempory proof theory knows what an essential role such
changes of order play in establishing various normal forms of first-order
proofs and thereby establishing many of the most central results of proof
theory.

Small wonder, therefore, if Pappus found the situation confusing and
in particular did not see how a 'resolution' could be carried out in a way
which would not anticipate the whole synthesis. We conjecture that this
intrinsic difficulty in carrying out the 'resolution' of a theoretical analysis

explains better than anything else the relative insignificance of such 'resolutions' in Pappus' practice.

The problematical 'resolutions' were saved from this fate by a body of special results (of the kind described earlier in section 1) concerning interdependencies of different 'given' elements.

11. THE AMBIVALENCE OF ANCIENT GEOMETERS TOWARDS AUXILIARY CONSTRUCTIONS

Although the historical material is too scanty to offer much direct evidence, our hypothesis squares in any case well with the general uneasiness of ancient geometers *vis-à-vis* the role and justification of auxiliary constructions. In particular, the need of constructions in the proofs of theorems puzzled them. Proclus, for instance, when faced with this 'anomaly', finds only an eminently unsatisfactory (because false) 'explanation': that a situation of this kind is exceptional (cf. Friedlein p. 204; Morrow, p. 159). In analytically conducted arguments the situation of this kind was even more irritating, for the reasons discussed above. No wonder, then, if it was in connection with analytical arguments that Heron apparently tried to remedy the situation by using semi-algebraic methods (cf. Ch. III, note 4, and Ch. VIII). By trying to dismiss auxiliary constructions altogether (even if he did not really succeed in doing so) he attempted to eliminate the trouble in one fell swoop. By the same token, of course, the 'resolution' lost its usual task; and in fact the Heronian 'resolutions' seem to have been of the rudimentary kind we discussed above. In the case of problems, Heron says, this approach is impossible. (Cf. *Anaritius*, ed. by Curtze, referred to in Ch. III, note 4. See p. 106, line 11ff.) But it is often impossible in analysing theorems, too. (For instance, Heron himself failed to deal with *Elements* 2, 1.) Hence it seems that what prompted his attempt to eliminate constructions in theoretical analysis was not so much the fact that it was possible to dispense with them (the scope of his methods is in fact not very large) but rather his general uneasiness with the constructive steps in the theoretical analysis.

The problem of understanding ancient 'resolutions' (in particular the question whether a 'resolution' was properly part of analysis or not) is thus a vivid illustration of the ambivalent position of auxiliary constructions in an analytically organized proof, i.e., in a theoretical analysis.

12. THE NATURE OF 'RESOLUTION'

It may be time to sum up what we want to say – at least tentatively – of the nature of 'resolution'. In Chapter V it was pointed out that when an analysis is inverted so as to become a synthesis, two different things have to be established: (1) the convertibility of each deductive step; (2) the independence of the auxiliary constructions of what has to be established (i.e., of the *zetoumenon*). It was surmised that the second task is what motivates the presence of 'resolution' in an analytical proof system. Even though our sample 'resolution' (above in section 2) shows that (2) is indeed what a 'resolution' accomplishes, it is clear that this cannot be the whole story. For it does not yet distinguish 'resolution' from the construction part of the synthesis nor indeed show why 'resolution' belongs to analysis and not to synthesis. This is not a real problem for us, however. For in Chapter IV it was pointed out that the very distinction between analysis and synthesis is due to a relative superficial, not to say artificial, aspect of the nature of the analytical method. It was due to the 'downward' direction of analysis proper in practice (cf. the single arrows in figures (5) and (6)). This in turn was occasioned by geometers' tacit decision to bring the information of the 'given' specified in K & A and of the *zetoumenon* specified in B to bear *in the same way*, that is, to use both of them jointly as premisses for logical inferences. Now this implies that whenever the distinction between the two directions of procedure becomes unclear, the distinction between analysis and synthesis likewise becomes unclear. But no matter how one looks at the direction of 'resolution', what it accomplishes (at least in the problematical case) is a step-by-step vindication of the auxiliary constructions carried out in the analysis proper *starting initially from the 'given'*, not from the *zetoumenon*. From the viewpoint of the desired conclusion or construction, 'resolution' thus proceeds upwards in the sense that the desired results are operating in it as conclusions, not as premisses. But if so, it is only to be expected that the boundary between the resolutive part of analysis and synthesis became unclear. The confusion which is undeniably present in the historical material therefore serves to confirm (albeit undirectly) our basic diagnosis of the whole situation.

In the course of our discussion we have nevertheless found a number of smaller characteristics that give 'resolution' something of an identity of its own. The most relevant (but still relatively unimportant) such characteristics seem to be the following:

(a) In the 'resolution', the order of constructions *vis-à-vis* deductive steps envisaged in the 'resolution' was usually the same as in the analysis proper, whereas in the synthesis all the auxiliary constructions were normally carried out before the proof proper.

(b) In a 'resolution', only the 'givenness' of the different constructions was established, i.e., the possibility of carrying them out somehow on the basis of the *dedomena* alone, independently of the *zetoumenon*, whereas in the constructive part of a synthesis the explicit construction was specified. Although this construction often did not contain anything really new, it summed up the so far scattered pieces of information in a convenient 'synthetical' formulation.

Neither of these features is very fundamental.

Why, then, was 'resolution' thought of as a part of analysis and not of synthesis? A partial but important answer is forthcoming as a corollary to our earlier answer (given in Chapter IV) to the question: What is an analysis analysis *of*? We said that analysis is essentially analysis of figures (configurations), not of arguments. In the light of this observation, an analysis is the more analytical the more exclusively it concentrates on anatomizing a definite figure. According to such lights, then, 'resolution' was the analytical part of a proof system *par excellence*, for no constructions were carried out in it but only the interdependencies of the objects of a fixed configuration were studied. With its nearly exclusive emphasis on the question of what is 'given', the Pappian 'resolution' offers a conspicuous example of the analysis of configurations, not of proofs. In this sense, the 'resolution' can even be said to be more analytical than the analysis proper, for constructions were usually carried out in the course of an analysis proper but never in the course of a 'resolution'. It is an ironical and instructive fact that it was in connection with this purest form of analytical procedure that the difference of analysis and synthesis and the direction of an analytical procedure became especially unclear.

NOTES

[1] See *Data* (cf. Ch. II, note 2), p. 252, line 23ff.
[2] Marinus' commentary can be found in Menge's edition of the *Data*, pp. 234–256. See p. 256.

[3] B. L. van der Waerden, *Science Awakening*, P. Noordhoff Ltd., Groningen, 1954, p. 198, note 1. Van der Waerden repeats the opinion of Dijksterhuis that 'determined' would be a better translation than 'given' for δοθέν. This translation is not wholly unproblematic, however. It is true that Apollonius (according to Marinus) had thought of the 'given' as 'determined' (τεταγμένον). But the adequacy of this view was denied by others (cf. *Data*, p. 234, line 15 and *passim*). Hence it is advisable to adopt the neutral translation 'given', which is also the literal one.

[4] See *Data*, p. 2 and Euclid's basic definitions there; *Data passim*.

[5] Friedlein, *Proclus*, p. 203, lines 5–6; *Data*, p. 236, line 1, and p. 248, line 11.

[6] *Data*, p. 248, lines 18–20. Cf. 248, 11ff.

[7] Heath, *Archimedes*, p. clxxxiii; cf. the 'resolutions' in the analyses at pp. 61–85, *op. cit.* (We do not try to distinguish here the contributions of Eutocius or of others in these propositions.)

[8] *Data*, p. 240, lines 24–25.

[9] *Data*, p. 250, lines 4–6.

[10] *Data*, p. 240 lines 9–10.

[11] There are gaps in the argument, as it stands in the Greek text. For a possible reconstruction, see Hultsch' translation.

[12] *Commentaires de Pappus et de Théon d'Alexandrie sur l'Almageste*, ed. by A. Rome, Vol. I (Studi e Testi 54), Bibliotheca Apostolica Vaticana, Rome, 1931, p. 35, line 22; p. 37, line 25; p. 34, lines 8–9. See also Hultsch 46, line 5 and 48, line 15.

[13] E.g., Hultsch 146, line 26. Comparing note 2 on p. 147 we see that also a construction is needed in that which is made "as in analysis" – thus the 'resolution' is ruled out.

[14] Hankel, *op. cit.* p. 141.

[15] Hankel, p. 147, line 27ff.

[16] A more correct characterization of the *diorismos* is offered (*pace* Hultsch 30, 14–16 where a description similar to that of Proclus is given) in Pappus' description of analysis (the passage is, according to Hultsch, an interpolation, however, and hence removed from our text and translation on p. 10): "*Diorismos* is a *preliminary* distinction (προδιαστολή) of when and how and in what cases the problem is possible." (Our italics, of course.)

[17] Heath, *Archimedes*, p. 68, line 1; *Elements*, Vol. I, p. 131, note 1, line 3.

[18] Pappus does not explicitly indicate the nature of these propositions. But they are theorems: not a construction but a proof is required in the enunciations (see, e.g., Hultsch 1006, line 3). Hultsch speaks of the main proposition among them as a theorem (1005, note*); similarly Heath in *Archimedes* p. lxiv, line 16, and p. lxv, line 23; and likewise M. Chasles on p. 44 of his *Aperçu historique sur l'origine et le développement des méthodes en géométrie*, Gauthier-Villars, Paris 1875 (second edition).

[19] *Comm. de Pappus*, p. 34, line 12; p. 37, line 25. Cf. p. 34, lines 8–9. – Ptolemy does not divide the original into theorems (cf. *Ptolemaeus*, ed. by Heiberg, Vol. I, Leipzig 1898, p. 370, line 4ff.). Many of the propositions of the commentary are really problems (see the comment of Rome at p. XX, *op. cit.*); but the proposition beginning at p. 35, line 12, is rather a theorem. At least this passage shows that for Pappus the phrase 'we analyse a theorem' when referring to an analysis with the 'given' terminology was not a solecism.

[20] See Niebel (*op. cit.* in Ch. I, note 12) p. 34; Friedlein, *Proclus*, p. 80; Morrow p. 66.

[21] Hultsch 30, lines 5–7.

[22] Friedlein, *Proclus*, p. 77; Morrow p. 63ff. Additional evidence can be found in M. C. P. Schmidt, 'Die Fragmente des Mathematikers Menaechmos', *Philologus* **42**

(1884), 72–81. Scattered remarks on the later history of this controversy are found in Proclus.

[23] Not even Speusippus thought that no constructions are needed in geometrical reasoning. Cf. Friedlein, *Proclus*, p. 179; Morrow p. 141.

[24] Friedlein, *Proclus*, p. 78, line 18ff.; Morrow p. 64, line 23ff.

[25] Hultsch 908, 20ff.; 910, 15ff.; 916, 25ff. Cf. also the theoretical analyses in the scholia on Euclid's *Elements* (see *op. cit.* in Ch. III, note 4: p. 366, line 23; 370, 7; 372, 15; 374, 9; 376, 10). ἔστιν δέ invariably also there in the 'resolutions' (of sorts); then a single step as in the *apodeixis* (but sometimes even this is missing). Cf. Ch. III, note 4, above.

CHAPTER VII

ANALYSIS AS ANALYSIS OF FIGURES: PAPPUS' TERMINOLOGY AND HIS PRACTICE

In the preceding chapters, an interpretation of the method of analysis and synthesis, as explained and practised by Pappus, was outlined. Some of the reasons which have led us to prefer this interpretation were only hinted at there. However, our interpretation can be backed up by a more detailed study of the evidence. Hence our remaining task is to indicate some of the further grounds for accepting this interpretation.

First, an evaluation of the available evidence is in order. The evidence concerning the method of analysis and synthesis in Pappus is of two sorts. On the one hand, there are the actual applications of the method (notably in the *Collectio*). Because there are tens of analytical arguments in Pappus, this evidence must be considered the decisive one. On the other hand, there is Pappus' general description of the method (see the text and translation in Chapter II, and Table I in Chapter III). It is reasonable to expect that these two pieces of evidence are compatible. If someone uses a method and also describes it, we are in most cases entitled to assume that he practises what he preaches. We shall attempt to see to what extent this is true of Pappus.

In evaluating the evidence, it is vital to understand Pappus' terminology. Unfortunately, it appears that Pappus' own surviving works are too narrow a basis for this task. Thus we are led to ask: What – besides the *Collectio* – is the relevant material in studying Pappus' terminology? Caution is advisable here, it seems to us. We may accept the testimony of the *Data*. (Pappus uses this work always in his problematical analyses; he mentions it at Hultsch 636, line 19; he had even written a commentary on it.[1]) We may accept Marinus' commentary on the *Data*. (Marinus refers to Pappus' commentary in the work of his own, and consequently knew it.[2]) We may also accept (besides the *Data*) the rest of the testimony of the great Alexandrian mathematicians from Euclid onwards. A thorough knowledge of this earlier work on the part of Pappus is witnessed by almost every page of the *Collectio*. We can safely presuppose that a certain influence was exerted by it also on Pappus' use of his terms. Then we have

Proclus, of course. Being a late writer, and in some questions the only witness, he must be included. But because he is a representative of a particular philosophical tradition (the Platonic one), certain caution is necessary.

On the other hand, Plato, and Aristotle, and their commentators must be largely excluded. This does not mean that we deny the existence of a tradition of analysis or Pappus' adherence to such a tradition. It means that we do not preclude an intervening development either in the method of analysis or in the use of the terms describing its several aspects. A considerable amount of scholarly work has been done on the early Greek mathematical terminology.[3] Because no mathematical works of that time have survived, such work relies mostly on Plato and Aristotle. For our purposes its results can be useful, too, but only as circumstantial evidence. As the first step, we must restrict our attention to Pappus and other late mathematicians. Only after we have appreciated what Pappus says can we compare his usage with the results of previous studies of the early terminology. This investigation falls mainly outside the present study; suffice it to say that blind reliance on the excluded material would in fact have been dangerous and misleading.[4]

The main connection between our interpretation of analysis and Pappus' terminology is the following. We have emphasized the possibility of viewing analysis as analysis of figures. Although the individual steps of analysis can be couched in deductive terms, i.e., as steps from one *proposition* to another, it is heuristically more suggestive to consider them as steps from a geometrical object (or a number of objects) in a certain configuration to a different object, in the sense of uncovering a connection between the former and the latter. (This interpretation is not spoiled by the fact that the connection can be expressed in propositional terms. In Chapters IV–V we saw what the logic of geometrical analysis is, and how an 'analysis of figures' can hang together with the corresponding formal proof.) This suggests that Pappus' terminology will be to a large extent one of geometrical objects and their interrelations and not a terminology of propositions and deductive connections. This expectation will in fact be fulfilled to a considerable extent.

Among other things, the terms describing the starting-points and end-points of analysis may be expected to be partly of this sort, that is, to refer to geometrical objects and their configurations rather than propositions. This will in fact be borne out nicely by an examination of such key terms of ancient geometry as the 'given' (δοθέν and the *dedomena*, δεδομένα; cf. Ch. VI above), the 'starting-points' (ἀρχαί), the 'things sought' or the *zetoumena* (ζητούμενα), etc. Most of this chapter is devoted to an examination of these notions.

We said earlier that many students of geometrical analysis have not seen what is really being analysed in an analysis. Similarly, sometimes it has not been seen clearly what Pappus is speaking of in his description of the method. His account nevertheless yields some telling clues when compared with his practice.

A major difficulty in appreciating Pappus' description and in connecting it with the actual geometrical practice is a certain tension there seems to be between the overwhelmingly 'constructional' vocabulary of Pappus' description and the possibility of correctly viewing his practice in propositional terms (see above). Hence an examination of the ancient views concerning geometrical *proofs* is in order here. This helps us also to understand better the internal structure of Pappus' description.

The theorems and problems of the ancients were general propositions (cf. Ch. III). But the proof was carried out in terms of a particular figure (and thus dealt with an instantiated form of the proposition). Proclus speaks of this 'instantial' way of proofs in so many words, and connects it with the use of a particular figure (not necessarily a *visible* figure).[5] On this view, the several steps of a proof were considered, not so much as steps between propositions, but rather as steps between geometrical objects with certain properties (based on their mutual interdependencies).[6] σύμπτωμα seems to be a name for a property of this kind; it can be found also in Pappus.[7] This view helps us to highlight the role of constructions, for a proof will not succeed without a sufficient number of auxiliary constructions. As a consequence even the whole proof was sometimes called a construction. Some of the supporters of this view went so far as to regard all propositions of geometry as problems (cf. p. 60 above). In view of the usual tendency of the ancients to group the deductive steps of proof into a separate compartment (thus removing constructions from this part of an argument), such as the *apodeixis* and to some extent also the 'resolution' in an analytical proof system, this view might seem a little exaggerated. In these deductive parts of an argument, at least, the 'constructional' aspect is in any case missing, one is tempted to say. However, this conclusion is too hasty. Many of these apparently purely deductive steps presuppose constructions in the sense that the proofs of those theorems which justify these deductive steps cannot be carried out without auxiliary constructions. This observation pertains also to the proofs of the theorems of the *Data* used in the 'resolutions' of a problematical analysis, not to speak of other deductive steps. (Cf. p. 57 above.)

That this way of viewing proofs was not altogether strange to Pappus can be seen already from his terminology. Following the practice of earlier writers, he speaks of *proving* as 'describing' (γράφειν), of *proving a lemma* as 'prescribing' (προγράφειν), and of a *theorem* as διάγραμμα (literally 'figure').[8]

Another indication of Pappus' adherence to this view can be seen from his general account on synthesis in Hultsch 634, line 22. Pappus speaks there of the construction of the thing sought in the synthesis. The thing sought (the latter part of an enunciation) of a theorem is a proposition (by modern standards). We do not construct propositions in the usual sense of constructing.

Pappus' practice of analysis, especially the presence of the 'resolution' with its emphasis on constructions and their properties, offers another major piece of evidence for our view, as we saw in Ch. VI. The fact that the proof was thus viewed by Pappus explains how he can speak of something which *also* is a series of deductions (e.g., analysis proper) in terms of constructions (and their interdependencies). Hence the tension mentioned on p. 72 above is not a real obstacle to our interpretation. This explanation helps us to relate in a natural way our subsequent discussion of the analysis proper with the observations we have made in the preceding chapters (especially in Ch. V).

Analysis begins, says Pappus, "from what is sought – as if it were admitted" (Hultsch 634, 11). The thing in question can apparently be *true* or *false* (636, 2; 5; 7), but it can also be *done* (634, 20), *be* (636, 1) and *be constructed* (634, 22).

Proclus says that τὸ ζητούμενον ('the thing sought') is the second part of a general enunciation (see p. 43).[9] This sense can be found in Pappus. For instance, at Hultsch 654, 19ff., Pappus says that "one must distinguish the kinds of these [porisms] not according to the hypotheses but according to the ζητούμενα...". Then (654, 25): "... the kinds of the *zetoumena* in the enunciations must be arranged as follows...". Pappus puts forward a common enunciation of a whole series of porisms. However, only the consequents (*zetoumena*, comparable with I(b) in our enumeration) are listed. Pappus can do this because the first part (which is explicitly given only in the first enunciation, the enunciation being the only complete one) is common to all these porisms. These enunciations are clearly *general* ones (cf. p. 27), not already instantiated enunciations.

In a sense, the force of the word *zetoumenon* varies somewhat in Pappus. For instance, at Hultsch 38, 23 it apparently means 'the solution sought for'. At 30, 3 it refers to a proposition. This kind of variety of meaning is easy to explain, however. In solving a problem (say) the final construction (sought for) accomplishes the whole construction and gives the solution to the problem at the same time.

In an analysis, the *zetoumenon* can be also the latter part of the 'instantiated' enunciation. When this happens in an analysis proper, later in the same analysis the verb ζητεῖν can also refer to a consequence of assuming the thing sought (ἐζήτουν: "because the thing sought is...", ζητεῖται: "on the basis of the thing sought...", Hultsch 204, 13 and 20). (Notice how neatly this usage fits in with our analysis of the structure of analysis in Chapter IV above.)

Another term for the starting-point of analysis is προταθέν (Hultsch 634, 26; 636, 8, 12). This word is used in Pappus' description of analysis and synthesis only when speaking about problems. The meaning of this term, too, seems to vary in Pappus. At 48, 13, for instance, the meaning seems to be 'the solution sought for'. At 650, 17 the thing referred to by the word is said to be an object of construction. (Whether the construction of the crucial lines – cf. our I(b) – or the whole construction is what Pappus has in mind, is not clear.)

The phrase προτείνω πρόβλημα ('propound a problem' at Hultsch 634, 4–5) might seem to suggest, however, that in Pappus' description of analysis προταθέν (from προτείνω) is a problem. But in the general enunciations of problems there is typically *only* 'the thing sought' (says Proclus, see Friedlein p. 204). Thus this term which in Pappus' description has the same function as the *zetoumenon* can well refer to the Part I(b) of the enunciation.

In practice, what is admitted first in analysis is not the general theorem or problem. What is admitted is an instantiated version of the conjunction of the antecedent and the consequent of the general enunciation. Because analysis differs from the usual (synthetical) procedure with regard to the *zetoumenon* (in assuming it), this *zetoumenon* too ('as admitted') can conveniently be called a starting-point. If we read the term like this, as we have reasons to do, the logical structure of the analysis proper is not, strictly speaking, described by Pappus, because in an analysis proper both the *dedomena* and the *zetoumenon* are starting-points of deduction (cf. pp. 36 and 41 above). Hence the propositional interpretation of analysis seems to fail here.

The information Pappus supplies concerning the end-points of analysis (analysis proper) is even more favorable to our analysis-of-figures view. The end-point of analysis proper (cf. Table I, p. 23, rows 5–6) is in the original something γνωριζομένων ἢ τάξιν ἀρχῆς ἐχόντων (Hultsch 634, lines 16–17). The formulation τάξιν ἀρχῆς ἐχόντων has usually been interpreted as meaning something like 'ranking as first principle'. Commentators have seen here axioms and postulates. However, *arkhe* is a notoriously loose word in Greek mathematical usage, a word which in different contexts has a great variety of different meanings (and similarly in philosophical language). In general, it can be translated as 'beginning' or 'basis' and it applies to a multitude of situations.[10] Pappus uses *arkhe* typically, although not quite exclusively, of mathematical objects and their interrelations, not of mathematical propositions.

In the *Collectio*, *arkhe* is mostly something that is given originally in the enunciation (*viz.* in the given hypotheses or in our I(a)). This is shown, e.g., by Hultsch 254, 19ff. where we read: "This does not follow from what is posited initially (οὐχ ἔπεται ταῖς ὑποκειμέναις ἀρχαῖς)... if this *ratio* is not *given*" (our italics). The missing *arkhe* is a *ratio*, and it is missing from the *hypotheses* (or τὰ ὑποκείμενα) of the enunciation (cf. ὑπόθεσις at 254, 2). Now the *hypotheses*, or the *hypothesis* in the collective sense, is that which is given originally, either in the *dedomena*-part of an enunciation (this is the case here) or in the premisses of a step of proof.[11] We see that the *hypotheses* were considered as consisting of same type of material as the 'given' (that is, not of propositions but, e.g., lines, points, ratios, etc.). The same observation can be made often elsewhere in Greek mathematical texts.[12] Thus the translation 'first principle' is often quite misleading.

Notice that there are in any case not inconsiderable difficulties in trying to interpret the *arkhai* of Pappus' general description of analysis as 'first principles' in the sense of axioms. For instance, if analysis is thought of as a downward movement or as a series of equivalences, it cannot always terminate in an axiom. For usually no one axiom is implied by the desired theorem which, being hopefully a logical consequence of the axioms, is typically weaker than they.

The term *arkhe* has in Pappus of course also other, nontechnical uses. For instance, at Hultsch 654, line 18ff. he says of Euclid: "He made only the beginning of this study" (ἀρχὴν τάξαι). Here τάσσω, on the other hand, means literally 'to make in certain order'. (Cf. also τάξαντες, at 634, 21.) Its use with ἀρχή shows that the corresponding noun τάξις at 634, 16–17 can well mean order. Elsewhere in Pappus' description of the 'Treasury of Analysis', moreover, τάξις really means order (636, 18).

At Hultsch 808, line 12 we find the similar use: "... in that which is originally posited in the problem" (ἐν τῷ ἐξ ἀρχῆς ὑπόκειται προβλήματι). Here the 'beginning' is an inequality. (This problem, however, seems to be an interpolation into another problem. Thus its hypotheses are not to be sought in the genuinely Pappian enunciation at 806, 24ff., but at 808, 4ff. Cf. Hultsch 808, *apparatus criticus*.)

'The first' (τὸ πρῶτον) is often used for *arkhe*. In logical contexts it often means, like the latter, 'indemonstrable proposition' (e.g., Aristotle's *Topics* I, 1, 100b27). But at Hultsch 174, 40ff. we read: "... the conclusion is the same but that which is *first* and by which it is obtained, is not. In the mechanical construction the conclusion is obtained when the *ratio* of two unequal straight lines is *given*, in the proof when the *ratio* is not *given*".[13] (Our italics.) Two separate arguments are being compared here, and the 'first principle' which the other has and the other has not among its *dedomena* is again a ratio.

In Marinus there is a sentence which is like a collection of the previously treated words. It is a definition of 'given': "... given is that which we can produce by means of that which is *posited* for us in the *first hypotheses* and *principles*" (κειμένων... ἐν ταῖς πρώταις ὑποθέσεσι καὶ ἀρχαῖς). Here the *hypotheses* are again to be understood as the *dedomena* (Menge's Latin translation catches the meaning of the sentence correctly. He even has *a nobis*, 'by us' for our 'for us'.)[14] Another passage in Marinus has an example of *arkhe* in similar use (cf. pp. 50–51 of our study).

As for γνωριζόμενον (Hultsch 634, 16; see also γνωσθέν at 636, 8) which is usually interpreted as referring to previously known theorems and problems, we have Marinus' report that the 'given' (δοθέν) had often been defined in terms of 'known'.[15] Hence we can expect that the 'known' is applied to the same kind of things as the 'given'. Concerning 'that which was seized last in analysis' (καταληφθέν, 634, 19), which is a still further expression for end-point, we likewise have the report that the 'given' was defined also in terms of καταληπτόν ('which can be seized in mind', *viz.* in the sense of the mind's comprehending it; see *Data*, p. 234, lines 12–13).

Thus it seems that the end-point of analysis is the *dedomenon*. Accepting this, we have again the ambiguity in the 'given'. Here the former part of the *general* enunciation seems to be ruled out, because in many problems no such part is present (example: 'describe a spiral').

The word ὁμολογούμενον (Hultsch 634, 12–13; 636, 3–4; 6; 10; 13) can also be applied to geometrical objects (cf., e.g., the scholion at *Data* p. 264, lines 5–6). But there is apparently a difficulty.

Certain passages in Pappus' general discussion of analysis, in particular in Hultsch 636, lines 4–7 and 10–14, show that instead of a *dedomenon* the end-point of an analysis might be an impossibility. Someone might be inclined to see here an assumption to the effect that analysis proper proceeds 'downwards' in Pappus' practice. Then the outcome of the analysis proper would be a *reductio ad absurdum* argument. In Proclus analysis is in fact related to proofs by *reductio* (Friedlein, p. 255). As was pointed out above in Chapter IV (pp. 36–37), we need not deny that in practice Pappus proceeded in his analyses 'downstream'. However, it seems to us remarkable that in his general description of analysis and its varieties reductive proofs do not merit more than a passing remark on the possibility of a 'negative' outcome. This lends perhaps some further support to our interpretation according to which Pappus' general characterization of analysis describes it as an upward movement. The references to an impossible outcome might then be interpreted in the light of the plain but sadly neglected fact that even the search for antecedents from which a given theorem could be proved can run into a dead end. In principle, such a contingency need not even result in a reductive proof, as a closer logical examination of the situation shows. (That is, the only legitimate conclusion might be a *non sequitur* instead of a reductive proof.)

We have to be careful here, however, for the neglect of proofs by *reductio* in Pappus' general description of analysis might have other explanations. First, there seems to have existed something of a tradition of analysis in which the negative outcome was habitually linked to proofs by *reductio*. To it belongs already Plato's description of *diorismos* (*Meno* 87A–B), and it continues all the way down to Pappus (see also his remarks on *diorismos* at Hultsch 30, 14–17) and to Proclus (see also Friedlein, p. 66, 22ff. where the *diorismos* is treated). This whole tradition is characterized by a tendency to speak about 'possible' *and* 'impossible' in contexts where either 'possible' or else 'impossible' alone would have been the correct expression. Pappus may simply have followed this slightly confused tradition. The confusion of this tradition is neatly betrayed by the fact that while the 'positive' outcome of an analysis (if successfully converted) proves a certain theorem S, the corresponding 'negative' outcome would prove its negation $\sim S$ (from the given premisses). Now often *neither* S nor $\sim S$ is provable from these premisses, wherefore an analysis must in any case be able to admit other types of outcome as well.

Secondly, and more importantly, the neglect of a *reductio ad absurdum* proof which may result from an attempted analysis may be partly explainable through the fact that such a proof is strictly speaking an undesirable outcome of the analysis proper. The analysis is said to be an attempt to prove a given (putative) theorem S (or to carry out a definite construction C). Now a reductive proof, conceived of along the traditional lines, would prove, not S, but not-S (or would prove, not the possibility, but rather the impossibility of C). Thus the 'negative' outcome of an attempted analysis might simply amount to something that does not any longer fit into the framework of an analytical proof system, and consequently was not pressed by Pappus into it. It is characteristic that instead of speaking of 'negative practical syllogisms' in his description of deliberation Aristotle says instead simply: "And if we come to an impossibility, we give up the search" (*Eth. Nic.* III, 3, 1112b25). In a sense, a negative outcome of an attempted analysis would hence result in something that is not an analysis any more. No wonder, therefore, that there for instance is no counterpart to the 'resolution' in a reductive proof.

Thus there does not seem to be anything that would absolutely force us to disclaim the tentative interpretation offered in Chapter II of Pappian analysis as an 'upward' movement. However, a fair amount of confusion is by any token present in Pappus. (More evidence to this effect will be offered soon.) It has already been seen that Pappus' general discussion, which gives rise to this problem, is only an imperfect representation of the much more intricate structure of actual analyses, as they occur in mathematical practice. (See Ch. IV and V above.) In the light of this more complicated structure, Pappus' statement concerning the abortive analyses becomes perfectly understandable.

Some further light on this matter is shed by an interesting 'resolution' of an impossible problem which is actually found in Pappus. (See Hultsch 30ff.) Pappus speaks of certain constructions in the *dedomena*-part as (incorrectly) admitted (ὁμολογούμενον), and moves from their impossibility to the impossibility of the whole problem. The general problem is 'how to find by plane methods[16] two mean proportionals in continuous proportion' (Hultsch 30, 25). The 'auxiliary' enunciation where a particular construction is proposed (by a colleague of Pappus) is found at 32, 4–26 (the *zetoumenon* is formulated at 32, 26). Pappus says repeatedly that the colleague takes "the thing sought for as admitted" (ὁμολογούμενον). He also says that the other geometer is deluted by the impossibility in the *beginning* (τὸ ἐξ ἀρχῆς ἄπορον, that is, by the impossibility in constructing the *dedomena*), and that he *constructs incorrectly* (ψευδογραφεῖ; the word can also mean 'proves incorrectly').[17]

The upshot of Pappus' criticism is that the proposed construction begs the question. The sought result is not reached (by 'plane methods'), because the constructions in the *dedomena* (apparently including the *zetoumenon* and an auxiliary construction) are impossible by 'plane' methods.[18] Now Pappus thinks that he can say that the thing sought is impossible also in the general case, i.e., independently of the particular auxiliary construction employed.

But Pappus has not proved that the general problem is impossible (perhaps he does not even claim that). At most, he has succeeded in showing that this particular construction, as it stands, does not yet solve the problem (see above p. 57).[19] Pappus is apparently confident of his case, but only because he believes that the problem is impossible by 'nature'. His proof of the impossibility thus reduces to a mere observation that the particular proposed figure does not yet solve the problem. In strict logical terms, Pappus' 'proof' of impossibility is thus faulty. Even if the constructions proposed in the *dedomena* are impossible (and Pappus has not even proved as much), it does not follow that the problem is impossible.

This (admittedly unique) treatment of questions of impossibility is interesting for two reasons. First, it supports our conjecture that Pappus' general statements about the 'negative' outcome of analysis (and hence also the 'positive' ones, of course) refer also to the 'resolution'. Moreover, the example with its begging-the-question idea squares nicely with the 'upward movement' insight suggested by Pappus' description. It is the impossibility 'in the beginning' which seems to make the problem impossible, and not an impossibility at the end of a deduction, as the 'downward' interpretation of Pappus' 'negative' statements would lead us to expect.

Secondly, on our interpretation the 'negative' statements (at 636, 6–7 and 13–14) were seen to be incorrect. But they are no more and no less incorrect than the example of Pappus' practice just discussed. We suggest that in both cases Pappus simply errs (or, at least, is unduly vague).

Thus we need not reject our previous conclusions about the description Pappus gives us. Analysis, so it seems, is described consistently as an 'upward' movement. As a description of the direction of actual analyses such an account is demonstrably incorrect and *very* imperfect.

The following fact is also relevant here. The end-points of analysis proper are apparently among the *dedomena*. But the last *proposition* reached in practice is never a proposition listed in the *dedomena* (cf. p. 24, Step$_{11}$; and p. 53). The step from the *dedomena* (the deductive one) to this last proposition is made in the *apodeixis*; the reverse on this deductive step is tried nowhere in the analytical proof system. What the analyst is in the first place interested in here (in the last step of analysis proper) is to establish a connection between the given geometrical entities and the auxiliary ones. The primacy of this interest is shown by the way in which the immediately following 'resolution' proves that the new entities and their properties are *really* given. A typical example of how the connection between things known and unknown is created is the end of analysis proper in our sample problem (p. 53 above).[20] The last proposition is an equation, and what is 'given' is the ratio whose square is the other side of the equation. This may be generalized. Quite often the 'given' entities arrived at in the last step of analysis proper are magnitudes, or ratios, whose squares, or products, or sums, or differences, or ratios are the other side of an equation which is the last proposition stated in the analysis proper. Then it is easy, of course, to prove in the 'resolution' that something on the other side of equation is given.

Now we are also in a better position to understand Pappus' description of the intermediate steps in the analysis proper. Although it can be conceived as a deduction starting from the *dedomena* and the *zetoumenon*-part of the enunciation, we have seen that Pappus' terms are not predicated on such a 'propositional' or 'deductive' view of the argument in question. That the word τὸ ἀκόλουθον does not describe the intermediate steps in such a deductive argument, either, was in effect shown in Chapter II. The precise sense of the word need not be sought for here, because the term has no such sense. This word is particularly handy in describing an analysis conceived of as an analysis of figures just because the various aspects of the process can be referred to only by a term which is not very precise.

Concerning the various aspects of the intermediate steps of analysis which are referred to by the expression διὰ τῶν ἑξῆς ἀκολούθων in Pappus' account on analysis, we will offer some remarks on one of them only.

Because the auxiliary constructions are in practice introduced at intermediate stages of analysis proper, they certainly belong to the things described. We recall that the cognate verb συνακολουθεῖν at Hultsch 252, line 10, meant a geometrical object's following another in certain way in a construction (similarly with ἀκόλουθον at Hultsch 34, 14–15; cf. Heiberg *Archimedes*, Vol. III, p. 172, lines 28–29).

On the other hand, the grammatical construction of the quoted phrase is instructive. In the language of the mathematicians (in Pappus and elsewhere) διά + gen. is typically attached to a word expressing something *by means* of which something is accomplished (e.g., proved).[21] Thus διὰ τῆς κατασκευῆς means 'by construction', and the same holds of particular auxiliary constructions (e.g., δι' εὐθείας, 'by straight line'; Hultsch 270, 5).

More specifically, a proposition can be said in Pappus' usage *to be analysed by means of* an auxiliary construction. At Hultsch 262, 24 we read: "... the problem can be analysed by means of a spiral..." (διὰ τῆς ἕλικος ἀναλύεσθαι), and at 258, 4–5: "... the problem can be analysed by means of surface-loci..." (διὰ τῶν πρὸς ἐπιφανείαις τόπων ἀναλύεσθαι δύναται).[22] The thing sought in both problems is the construction of the *quadratrix* (a curve by means of which the circle could in turn be 'squared'). Once again, this illustrates vividly the importance of auxiliary constructions for the method of analysis.

The word ἀκόλουθον is explicitly used for auxiliary constructions in a fragment of Speusippus *apud* Proclus.[23] A postulate (Proclus' examples are Post. 1 and 3 in Euclid) and a problem are compared. The former is ἀκατάσκευος, 'free of construction', whereas the latter is not. It is this difference Proclus illustrates by the quotation from Speusippus. Certain entities are reached immediately (being constructed by using postulates), others only by the ἀκόλουθον of these entities (i.e., by the auxiliary construction used in problems), according to Speusippus. (The examples mentioned here are problems: the construction of a spiral and of an equilateral triangle.)[24] Speusippus was also among those who thought that all propositions of geometry are theorems. Thus this fragment incidentally shows that even this view of geometry did not exclude auxiliary constructions.

All told, an examination of Pappus' terminology and practice thus offers further evidence for our 'analysis-of-figures' interpretation of the analytical procedure. It has also been found, in the light of this terminology, that Pappus' general description of analysis and synthesis is compatible with this interpretation. In particular, Pappus' practice shows that his statements concerning a negative outcome of an attempted analysis (cf. Ch. II, last few paragraphs) do not constitute an objection to our interpretation.

NOTES

[1] See *Data* (*op. cit.* in Ch. II, note 2), p. 256.
[2] *Data*, p. 256.
[3] See, e.g., B. Einarson, 'On Certain Mathematical Terms in Aristotle's Logic', *The American Journal of Philology* 57 (1936), 33–54 and 151–172; Ž. Marcovic, 'Les Mathématiques chez Platon et Aristote', *Bulletin International de l'Académie Yugoslave des Sciences et des Beaux-Arts, Cl. des Sciences Mathématiques et Naturelles* 32 (1939), 28–48; O. Becker, 'Die Archai der griechischen Mathematik', *Archiv für Begriffsgeschichte* 4 (1959), 210–226; K. von Fritz, 'Die APXAI in der griechischen Mathematik', *ibid.* 1 (1955), 13–103; Á. Szabó, *Anfänge der griechischen Mathematik*, R. Oldenbourg, Munich and Vienna, 1969, p. 361ff. See also P.-H. Michel, *De Pythagore a Euclide*, Société d'Edition 'Les Belles Lettres', Paris, 1950.
[4] Several of the authors mentioned in note 3 for instance try to establish the terms which in the early Greek geometry stood for such important concepts as 'axiom', 'theorem', etc. Suffice it to say that neither their results nor their arguments can be extended to Pappus, at least not without a great deal of further argument, – As for Pappus' relation to earlier Alexandrian mathematics, see B. L. van der Waerden, *Science Awakening* (*op. cit.* in Ch. VI, note 3), p. 267.
[5] Friedlein, p. 207; Morrow, p. 162. For figures imagined in mind only, see Friedlein, p. 78; Morrow, p. 64.

[6] Niebel (*op. cit.* in Ch. I, note 12), p. 35ff. The introduction of new properties and relations into a proof was apparently considered by the school of Menaechmus to be a kind of construction. Cf. Friedlein, p. 78; Morrow, p. 64.

[7] E.g., Hultsch 252, line 2.

[8] See Hultsch' index on these words. For the similar practice earlier, see Niebel, *op. cit.*, p. 92ff.

[9] Friedlein, p. 203; Morrow, p. 159ff.

[10] For some common ideas underlying the predilection of many Greek thinkers for 'principles', see E. W. Beth, *The Foundations of Mathematics*, second ed., North-Holland Publishing Company, Amsterdam, 1968, pp. 8–9 and pp. 31–32.

[11] Hultsch 254, 2; 665, 7. Cf. the commentary of Marinus, *Data*, p. 232, line 2.

[12] Heath, *Archimedes*, p. clxxxiii; *Elements*, Vol. I, p. 303, note 29. See also Hultsch 36, 4; 36, 23; 40, 23; 44, 9–10 and 11; 46, 12. Cf. Friedlein, p. 244 (Morrow, p. 190) where *hypothesis* = a member in Part I(a) of an enunciation.

[13] The text goes on: "... thus it remains to show, how to find four straight lines in given ratio". Someone might say that the proposition in the 'how'-clause is the 'principle' and that Pappus describes two different numbers of axioms and postulates used in the two methods. To this we can reply that Pappus speaks first of ratios; and the proposition in the 'how'-clause is in any case a problem, not a postulate.

[14] *Data*, p. 252, lines 9–10.

[15] *Data*, p. 234, line 17; pp. 234–257 *passim*.

[16] For 'plane' methods, see Heath, *Archimedes*, p. cxl; O. Becker, *Das mathematische Denken der Antike*, Vandenhoeck & Ruprecht, Göttingen, 1957, p. 74.

[17] Hultsch 38, 19; 40, 12; 44, 13; 46, 18. ψευδογραφεῖ at Hultsch 40, 17; ἀρχή at 40, 16.

[18] Hultsch 38, 17ff. Cf. the *dedomena*-part at 32, 12ff.

[19] See Thomas (*op. cit.* in Ch. II, note 3), Vol. II, p. 568, note *a*. The method does not solve the problem, but does furnish a series of successive approximations to the solution. – As Becker (*op. cit.*, note 16 above) p. 75, note 14 says, it is not clear whether such proofs of impossibility (for instance, impossibility by the so-called 'plane' methods) were carried out correctly in Antiquity at all. See also A. D. Steele, 'Über die Rolle von Zirkel und Lineal in der griechischen Mathematik, Teil II und III', in *Zur Geschichte der griechischen Mathematik*, ed. by O. Becker (Wege der Forschung, Vol. 33), Wissenschaftliche Buchgesellschaft, Darmstadt, 1965, pp. 146–202.

[20] The last proposition is '$AC^2 : CB^2 = AD : DB$'; '$AC : CB$' is given.

[21] Heath, *Archimedes*, p. clxxxiv.

[22] Hultsch translates in both cases *solvi*, 'solve', for ἀναλύεσθαι, probably because there is no separate synthesis in these problems. The method in them is analytical, however, and thus we can with VerEecke translate as we have done.

[23] *De Speusippi Academici Scriptis* (*Accedunt Fragmenta*), P. Lang, diss., Bonn, Typis Caroli Georgi Typographi Academici, Bonn, 1911, p. 29; p. 62, fr. 30 = Friedlein, *Proclus*, p. 179, line 8ff.

[24] The original text (we follow Lang) is as follows: καθόλου γάρ, φησὶν ὁ Σπεύσιππος, ὧν ἡ διάνοια τὴν θήραν ποιεῖται, τὰ μέν ... τὰ δὲ ἐκ τοῦ εὐθέως αἴρειν ἀδυνατοῦσα κατὰ μεταβάσιν ἐπ' ἐκεῖνα διαβαίνουσα κατὰ τὸ ἀκόλουθον αὐτῶν ἐπιχειρεῖ ποιεῖσθαι τὴν θήραν.

CHAPTER VIII

PAPPUS AND THE TRADITION OF
GEOMETRICAL ANALYSIS

In this chapter a brief survey of the tradition of geometrical analysis in Antiquity will be attempted. It will appear that although geometrical analysis was sometimes used as a methodological paradigm, frequently mathematicians' and philosophers' interpretation of the geometrical method was conversely influenced by their philosophical and methodological ideas[1].

Although Plato does not use the term 'analysis' either for a geometrical or for a philosophical method, some sources allege that he was the inventor of the very method of geometrical analysis.[2] Usually it is assumed, however, that the analytical method was employed already by Hippocrates of Chios and perhaps even by the early Pythagoreans.[3] Unfortunately, no undisputably genuine mathematical arguments have been preserved to us from those early times (before Aristotle). Eutocius, it is true, ascribes two analytical arguments (analyses of problems) to Menaechmus,[4] but they might have been carried out later by someone else, although certain features in them (such as the use of conic sections) are probably genuine.[5]

The rest of the extant evidence pertaining to mathematicians before Heron concerns almost exclusively the problematical kind of analysis. If Hippocrates and the Pythagoreans really knew the method, they probably applied it only to problems (such as the duplication of the cube, or the construction of the regular pentagon).[6] As we shall see, for Aristotle, too, the paradigmatic kind of analysis was the problematical one. In fact, Carpus of Antiochia (first or second century A.D.) *expressis verbis* connects the use of analysis with problems, not with theorems.[7] Admittedly, the analyses ascribed by Proclus to Eudoxus (Friedlein p. 67, line 2ff.) might or might not have been theoretical ones.[8] The analytical comparisons of regular solids mentioned at one point by Pappus (Hultsch 410) were probably theorems. But we do not know who the 'ancients' were who made these comparisons. Apparently even Heron (and consequently every one of Heron's contemporaries and predecessors) could be 'ancient' for Pappus (see Hultsch 54, line 3, – 56, line 1).

Owing to the absence of genuine material, we do not even know what the problematical analyses of Aristotle's time were like. In particular, we do not know whether the analyses Aristotle had in mind in the *Nicomachean Ethics* III, 3 (see below) contained the part we have called 'resolution'.

Many historians of mathematics seem to assume that the early form of analysis was reduction (*apagoge*).[9] But if so, this only helps us to reformulate the problem concerning the 'resolution' we face here. In reality the nature of reduction of a problem to another is virtually as problematical as that of its alleged analytical counterpart. Zeuthen has suggested, it is true, that the part of the (late) analytical proof system which we have dubbed 'analysis proper' (that is, analysis in the broader sense minus the 'resolution') was called 'reduction' (*apagoge*). Assuming that the early analysis amounted to *apagoge*, we might think that early analyses contained no 'resolution'.[10] However, Zeuthen's equation of the *apagoge* with our 'analysis proper' (of late Antiquity) is clearly wrong. There is plenty of evidence showing that when an analysis turned out to effect a reduction of the problem at hand to another problem, this typically happened in the 'resolution'.[11] The way in which this happened is as follows: certain parts of the proposed construction were first shown to be given (from the original data of the problem), and then it was recognized that the remainder was justifiable by a problem already solved.

Thus no real information concerning the nature of early analysis is forthcoming from the study of the late *apagoge*, either.

However, the presence of the usual means (similar to the propositions of Euclid's *Data*) of carrying out 'resolutions' already in Aristotle,[12]and the brevity of the timespan between Aristotle and the oldest writers (Aristaeus, Euclid) represented in the 'Treasury of Analysis', suggest that the 'resolution' had in any case been discovered even if it was not yet being generally used in Aristotle's time. But was the 'resolution' (if it was used in connection with analysis) then generally thought to be a part of analysis, as it was in Pappus' time? Did Aristotle have in mind analyses containing a 'resolution'? We simply do not know.

Be these questions as they may, Aristotle in any case knew the geometrical method of analysis perfectly well. It is referred to in Aristotle's famous discussion of deliberation (*bouleusis*) in the *Nicomachean Ethics* (III, 3, 1112b15ff.), and is compared by him there with deliberation (or 'planning').

In fact, this passage is reminiscent of segments of Pappus' description of the geometrical method to a truly astonishing extent. The astonishment is partly due to the fact that Pappus does not usually advocate any specifically Aristotelean doctrines; the name of the Stagirite does not even occur in the *Collectio*.[13] (Nor are Theophrastus or Eudemus mentioned, whereas Pappus sometimes speaks of 'divine' Plato.) This makes especially remarkable the fact that Pappus' description of analysis and synthesis and the Aristotelian passage describing *bouleusis* and comparing it with mathematical analysis at *Eth. Nic.* III, 3, 1112b15–29 are closely similar.

The vocabulary of the two passages already witnesses to this similarity. Cases in point are τέλος at 1112b6 and *telos* at Hultsch 634, 22; ἀρχή at 1112b28 and *arkhe* at 634, 17; σκοποῦσιν at 1112b16 and ἐπισκοποῦσιν at 1112b17 and σκοπούμεθα at 634, 14; εὑρέσει at 1112b19 and εὑρετικήν at 634, 6; ζητεῖν and ζήτησις at 1112b20 and 22 and ζητητικήν at 634, 24; ἐντύχωσιν at 1112b24–25 and ἐντύχωμεν at 636, 6–7 and 13–14; πορισθῆναι at 1112b26 and ποριστόν at 636, 11 and ποριστικόν at 634, 25.

What is even more important are the structural similarities. Both passages speak of a hypothetical starting-point of analysis (cf. θέμενοι at 1112b15). Both speak of a procedure in terms of its aim (*telos*). In both accounts the last thing in the procedure is the first thing in the opposite 'movement' (in the *synthesis*, or in the *genesis*, respectively).

In both, there is also a description of the two possible outcomes (of analysis or of *bouleusis*, respectively).[14] Last but not least, in both descriptions we seem to have an 'upward' movement. That analysis (be it geometrical or philosophical) was in general conceived of as an 'upward' movement before Pappus is argued for by Gulley (cf. Ch. II above). However, it is perhaps safer to say that it was not conceived of as a 'downward' movement before Pappus; sometimes the direction of analysis is not clearly indicated. We shall see certain examples of this below.

Thus Pappus' account can be seen to be strongly reminiscent of certain Peripatetic (or Aristotelian) formulations. Notice, however, what this similarity does not cover. There is no trace of Aristotelian syllogistical theory in Pappus' description.

The interpretation of analysis which Aristotle presupposes in the *Ethics* passage is nevertheless not entirely clear. Although Aristotle sometimes emphasizes the role of constructions in mathematical heuristics (cf. *Met.*

IX, 9, 1051a21), it is possible that in the *Ethics* he has instead an essentially propositional interpretation of analysis in mind. However, the passage is formulated in rather general terms. For instance, Aristotle does not say in so many words what the mathematical counterpart to the *telos* or *arkhe* of a *bouleusis* is (although his discussion of *bouleusis* as a special kind of *zetesis* strongly suggests that the *telos* is what a mathematician wants to establish, i.e., the desired result). The passage, especially when considered in isolation, can easily be understood in a different way, and hence adjusted with minor changes to the purposes of our analysis-of-figures view. The analysis Aristotle speaks of can be seen to be of the *problematical* kind (cf. *diagramma* at 1112b21, and compare Aristotle's statements about the outcome of analysis with Pappus' statements in the problematical case). Now it is difficult to speak of construction problems without speaking of constructs (geometrical objects). One can of course speak instead of the operations (constructions) that bring them about. But even then one usually has to say something about the objects these operations handle or bring about. Thus in Aristotle's passage the means (*organa*) by which the desired result or action is to be brought about are perhaps a partial counterpart to the constructs (auxiliary entities) of a geometrical analysis.

Geometrical analysis is referred to by Aristotle also at *Soph. El.* 16, 17a28. There Aristotle says that synthesis may fail even after a successful analysis (cf. p. 55). This observation shows that already for Aristotle analysis and synthesis were complementary methods, or perhaps better, two parts of a single method of analysis and synthesis. An interesting aside in *An. Post.* I, 12, 78a6ff., shows that the Stagirite was fully aware of the convertibility problem in analysis, and of the fact that establishing convertibility of the different steps of analysis is the main burden in justifying analysis synthetically. (Cf. Sir David Ross' comments on this passage.)

Certain other passages in Aristotle are relevant here. A case in point is just *Met.* IX, 9, 1051a23ff., where the role of auxiliary constructions in geometrical heuristics is emphasized, although the word *analysis* is not used in that passage.

The Aristotelian *Ethics* model was not lost sight of by commentators. Thus Alexander the Commentator (who apparently lived only a little earlier than Pappus) writes:

Analysis is the way from the goal (*telos*) to the beginning. The geometers are said to analyse when they begin from the conclusion and proceed to the things which are first and to the solution (?? *problema*), following the order of the assumptions which are made for the sake of proof. (*In An. Pr.* I in CAG II, 1, p. 7, lines 14–18.)

It is fairly obvious that the conclusion here is the proposition which is analysed. However, Alexander does not say what the *arkhai* of an analysis are, nor indeed what the auxiliary 'assumptions' are that are made for the sake of the proof. Furthermore, the direction of analysis is not clearly indicated by Alexander.

John Philoponus, who lived nearly two centuries after Pappus, is much more precise. His description of the geometrical analysis is as follows:

Analysis is the reverse of synthesis. Taking the original thing sought (*zetoumenon*) as granted, e.g., that this triangle is isosceles we inquire what the premisses (*protaseis*) are by which it can be established (*kataskeuasthe*), and then analyse it further in order to find them, until we arrive at something admitted or at the principles of geometry. (*In An. Post.* I in CAG XIII, 3, p. 162, line 22ff.)

Philoponus does not mention the *telos* of the procedure here, but it seems to be for him the same as for Alexander (cf. above),[17] that is, the conclusion of the argument in question. It is not mentioned here because the *zetoumenon* is not the proposition to be analysed for Philoponus, but rather a (sought-for) *term* in it.[18] The "principles of geometry" are (for Philoponus) points, (straight) lines, and so on.[19] Their definitions, so Philoponus implies in a passage of *In Phys.* p. 333, line 3ff. (CAG XVI), play an essential role in a geometrical analysis on his interpretation. (Of course, it is the definitions of kinds of geometrical objects that are involved here, not definitions of individual geometrical objects.)

On this interpretation, what is analysed in analysis can hardly be anything but connections between different terms. The paradigm of explanation presupposed by Philoponus seems to be the one offered by the Aristotelian theory of the syllogistic structure of science. (See Jaakko Hintikka, 'On the Ingredients of an Aristotelian Science', *Nous* **6** (1972), 55–69.) This implies that scientific propositions can be analysed into 'atomic' ones in which two terms are connected in such a way that no further term can be inserted between them. Not unexpectedly, we find this explanation as applied to a mathematical method in Philoponus. For Philoponus considers mathematics purely theoretical science.[20]

In curious contrast to this, we nevertheless can see that Philoponus interprets analysis as an analysis of a *particular* figure (of "this isosceles triangle"; see above). In fact, in the example he gives of analysis at *In Phys.* p. 333, 6ff., he moves step by step from one geometrical object (e.g., angle) to another, establishing all the time interdependencies ('is equal to', for instance) between these different objects. When he says that one thing 'follows' another (ἐφεξῆς, cf. ἐξῆς in Pappus' account) he is always speaking of geometrical objects which come after each other. Furthermore, certain auxiliary constructions are involved in his example 'the sum of the angles of the triangle is equal to two right angles'. The sample proposition which is analysed, although borrowed from the passage of Aristotle's *Physics* (II, 9, 200a17) Philoponus is commenting on, is the same as in *Met.* IX, 9. Hence Philoponus must have been aware not only of the need of constructions in geometrical analyses, but also of their heuristical role according to Aristotle. However, Philoponus carefully avoids this topic, and simply assumes that the auxiliary constructions have been carried out in advance.

Thus we might characterize Philoponus' position by saying that he knew well enough the nature of geometrical analysis. As a theoretician he nevertheless preferred the conventional 'scientific' and 'theoretical' model of Aristotelian science. How such an Aristotelian account could be made to fit the facts of analytical practice is nowhere spelled out by him.

Needless to say, the Peripatetic-looking elements in Pappus' description imply no such orthodox position. Pappus' description is less 'theoretical' than Philoponus' and also less Peripatetic. As far as the facts of analytical practice are concerned, however, Pappus and Philoponus agree more than might be expected; the differences lie on the level of explanation.

The Peripatetic-looking expressions in Pappus' general description of analysis and synthesis are not all there is to it. Among the ingredients that cannot easily be interpreted in terms of a Peripatetic tradition there are the passages where the word *akolouthon* (ἀκόλουθον) occurs.[21]

This word occurs (besides Pappus) in the ancient descriptions of geometrical analysis only in a passage interpolated into Euclid's *Elements* (with five theoretical analyses; cf. Ch. III, note 4).[22] This passage is corrupt, and we do not know its author or age.[24] It may be by a late hand, or it may be an old description of the geometrical method which has been made unintelligible by an error of some copist.

However, the word *akolouthon* and the cognate verb *akolouthein* appear in a description of analysis (ἀνάλυσις) as a general philosophical method. Albinus, a second century Platonist, describes three types of philosophical analysis in Plato. The first of them is the passage from sensible particulars to the Forms, whereas the second is an 'upward' search of prior conditions (cf. Gulley, 'Greek Geometrical Analysis', p. 6). The third is "analysis by hypothesis":

> Analysis by hypothesis is as follows.... The investigator assumes something. Then he tries to find something which goes together with (*akolouthei*) the assumed thing and after that, whether he can give an account (*logon apodidonai*) of the thing assumed (*hypotheseos*). Then, making another hypothesis, he investigates whether the first hypothesis is in turn connected (*akolouthon*) with the other one. He repeats this until he comes to a starting-point which is not hypothetical (*arkhen anypotheton*).[24]

This analysis is modelled after the 'upward way' of Plato's *Phaedo* and *Republic*.[25] (Even the word *akolouthein* – in the sense 'to follow' – is found in *Phaedo* 107B in connection with the hypothetical method. It is the investigator who 'follows' the different hypotheses in order there, however, unlike the hypotheses which 'follow' each other in reasoning in Albinus.)

What is the meaning of the word *akolouthon* here? In another passage Albinus uses *akolouthia* for implication, and tries to find anticipations of this Stoic connective of his own day in Plato.[26] It is just possible that we have another similar case here. If so, this interpretation of Plato (if it was intended as one by Albinus) is in reality not very Platonic. In Plato's hypothetical method, the relation between different hypotheses is expressed by the word συμφωνία, 'harmony'.[27] This word is not used by Plato only, or even primarily, for logical implication.

Similarly, there is no evidence for the assumption that before the formation of the Stoic concept of *akolouthia*, the word *akolouthon* in context of a geometrical argument would have meant 'consequence'. At least once (in Speusippus, cf. p. 82 above) it had in such a context an entirely different meaning. In view of such facts, this late passage in Albinus on a philosophical method does not show that in an old description of geometrical analysis (for instance, in the passage interpolated into Euclid) *akolouthon* should stand for (logical) consequence. The occurrence of the word in Albinus and in Speusippus might even be a kind of link with an early Academic tradition of geometrical analysis. This is quite conjectural, of course, but it finds some support, e.g., in the fact that the terms ἑξῆς, ὁμολογούμενον (both of them are found in Pappus and in the Euclidean interpolation), and even ἀρχή, can well be associated with Plato's accounts of hypothetical method.[28] But clearly there are also other explanations compatible with the evidence.

It is perhaps in order to say something at this point about the elements of Stoic logic in Pappus' description. There are none (with one possible exception: the term καταληφθέν at Hultsch 634, 19 could be Stoic; but the use of the word was also common enough among the rivals of the Stoics). On one hand, all references to construction (Hultsch 634, 22) or to problems are beyond the scope of the propositional logic of the Stoics. On the other hand, although the word *akolouthon* was apparently used by the Stoics in the sense of 'consequence', the proper technical term for the consequent of an implication was λῆγον, not *akolouthon* or *hepomenon* (Hultsch 634, 20).[29] And although the Stoics practised an 'analysis' of their own, nothing suggests that Pappus was following this model, either in terminology or otherwise. The Stoic 'analysis' was a propositional procedure: complex arguments were reduced by means of certain rules (*themata*) to the five basic indemonstrable arguments.[30] There seems to have prevailed not only a distinction but even an opposition between the two kinds of analysis, the geometrical and the Stoic one. For instance, we know that Galen wrote a treatise entitled 'That Geometrical Analysis Is Better Than the Stoic One'.[31] (Notice, however, that the Stoics – Posidonius, for instance – sometimes themselves wrote on subjects that went beyond the scope of their *propositional* logic. The Stoic 'analysis' was not an attempt to interpret the geometrical method but an independent logical discipline which originally emulated the Aristotelian 'analytics', that is, syllogistic. Thus, while denying the influence of Stoic propositional logic on Pappus, this claim is not incompatible with Stoic influences of another kind.)

There are thus different traditional elements present in Pappus' general description of analysis and synthesis. This does not yet mean that the description is inconsistent. Analysis as a philosophical method was in vogue in the centuries before Pappus. In fact, widely different methods were called 'analysis'.[32] There were also different descriptions of the geometrical method, and some of them were rather popular. (For instance, Plutarch knew the geometrical method; cf. *Vita Romuli*, Cap. 9; *An Seni Respublica Gerenda Sit*, Cap. 16.) None of these seem to have been satisfactory, or generally accepted. If Pappus used this pre-existing knowledge – and pre-existing modes of expression, borrowing from different sources – his procedure is very natural in the circumstances. Pappus' contemporaries and the thinkers of the immediately preceding centuries were wont to write and to philosophize in an 'eclectic' manner, and Pappus apparently only followed suit.

Although it would be interesting to know what the mathematicians before Pappus thought about the method of analysis, the evidence is not very extensive. Geminus the Stoic seems to have written about the method, or at least defined it as "a method of finding the proof".[33] More informative is the description of Heron *mechanicus*, which is accompanied by actual analyses (cf. Ch. III, note 4).

Heron's analytical work is partly preserved in an-Naīrīzī. The actual analyses he carries out square well with our interpretation of analysis and synthesis. As T. L. Heath says, the method in them seems to be "*splitting-up* rectangles and squares, and *combination* of them into others" (italics by Heath).[34] Although Heron proceeds "without the figure", his method probably is only a semialgebraic variant of the usual theoretical analysis (cf. Ch. III, note 4). These analyses also seem to contain a rudimentary 'resolution' which is repeated in the subsequent synthesis. At least the translator sometimes repeats the last step of analysis (*dissolutio* or *resolutio* in the Latin of Gherard) in the beginning of the synthesis (*compositio*), and sometimes even says it: "I begin from the place I arrived in the analysis" (Curtze *Anaritius*, p. 93, lines 23–25). This remark, and similar ones (see also *Codex Leidensis*) are almost certainly added by the translators. The possible changes in the text (by different translators) recommend caution, and a closer study of these analyses in general, and in particular of the apparent 'resolutions' they seem to contain, would be needed before any definite conclusions are drawn.[35]

The same must be said of the description of analysis and synthesis we find in Heron (*apud* an-Naīrīzī).[36] We believe, however, that there is one feature in Heron's description that is genuine. It contains an expression of the view that an endpoint of an analysis is "something proved earlier".[37]

This is shown by Gherard's Latin translation, and especially by the Arabic text of *Codex Leidensis* which seems to be more reliable than Gherard's translation.

The following translation from the Arabic is provided by Mr Tapani

Harviainen, Department of Oriental Languages, University of Helsinki:

We suppose the thing sought ('al-maṭlūb, ζητούμενον?) as being (mawğūd, ὡς ὄν?). Then we resolve it (nafuḍḍuhu, λύω? διαλύω?) to something already proved (burhān). (*Codex Leidensis* 399, 1, Vol. II, 1, p. 8, lines 11-12.)

The Arabic word here for 'proof' (burhān) is clearly here a counterpart for the Greek *apodeixis* or for a cognate of the *apodeixis* (see *Encyclopedia of Islam*, New Edition, Vol. I, pp. 1326–1327 on 'al-burhān', and Curtze *Anaritius*, p. 40, line 33, where Gherard uses the word *probatio* for the geometrical *apodeixis*. Because Gherard speaks also of the *probatio* of the endpoint of analysis in his translation of the description of analysis, the two texts support each other).

What is this "something proved earlier"? It clearly might be a theorem. This admission obviously need not affect our view of the logical structure of analysis. Of course earlier theorems have a role to play in analysis. In fact, we already said that they have one (cf. schema (5), p. 36 above). The question is, rather, what kind of role? It might be (and sometimes has been) assumed that the endpoint, being a theorem, is inferred from the proposition we want to prove (cf. p. 11 above). On our interpretation of the analytical practice, earlier theorems have at all stages of the analytical argument (proof system) the role of premisses. The Heronian analyses are not an exception in this respect.

Nor does Heron say that the endpoint is inferred from something; the translations suggest that the thing sought for is "resolved" (λύω? διαλύω?) to this endpoint. Hence Heron's view about the role of the endpoints seems to be compatible with our analysis on this interpretation.

In Heron's usage one of course 'proves' theorems. But the words for 'proof' and 'prove' have also other uses in Greek (or modern) mathematics. For instance, one can prove that a certain 'straight line is a tangent of a circle', although this is not a theorem, in the midst of a longer argument. This familiar usage is found in Heron.[38] It is possible that the endpoint is a statement of this sort: not a theorem but a statement which can be proved from certain assumptions (*dedomena*).

The third possibility is that Heron has in mind, not just a theorem but its 'exposed' (instantiated) version. All these interpretations make good sense of the passage.

Because we thus cannot say much of the endpoints in Heron's description, its relevance for our interpretation of Pappus' description remains partly open. If the thing already proved in Heron is a statement depending on the *dedomena*, this of course squares nicely with our interpretation of Pappus' description. Otherwise, the situation becomes more complicated.

In Pappus the words for endpoints which can cause trouble are 'beginning' (*arkhe*) and 'known' (*gnorizomenon*). Now Heron does not use these words. In Ch. VII we saw that these words can easily have other meanings than 'axiom' and 'previously known theorem' in Pappus. We can of course still stick to our interpretation of the import of these words in *Pappus'* description, in spite of the fact that for *Heron* the endpoints might be theorems. Furthermore, Pappus (although he probably knew also this particular passage of Heron) does not use the term 'previously proved' in his description of analysis, and the word *arkhe* (which he used) seems to have elsewhere in Heron uses reminiscent of Pappus' description of analysis (on our interpretation).[39] It is possible, however, that the phrase "as being" (Hultsch 636, 1 and 3) and Pappus' account of *theoretical* analysis in general, might be inspired by Heron whose work in other fields Pappus knew very well.

So far we have discussed such descriptions of geometrical analysis as exhibit at least a remote similarity to aspects of Pappus' account. As we have seen, the picture is far from uniform. But there is also evidence of other, different conceptions of geometrical analysis. Two of these competing conceptions are especially interesting. On one hand, analysis was conceived of as a path from a complex whole to its simple parts. If the complex starting-point is taken to be the configuration to be analysed (as specified by the enunciation of a proposition), this conception clearly comes very close to our 'analysis as analysis of figures'. On the other hand, geometrical analysis was on at least one occasion also interpreted as 'downward' propositional inference, comparable with a reductive argument. The latter interpretation of analysis is found among the different accounts on geometrical analysis in Proclus. This seems to be the only passage in which analysis is in so many words represented as normal deduction.

John Philoponus once speaks of geometrical analysis as a path "from a whole to its parts and from a complex to the simple".[40] No further explanations are given. In Proclus we sometimes find geometrical analysis conceived of as a division of a figure (geometrical configuration) into "simple" (*hapla*) parts.[41] This comes near to our analysis-of-figures view, and can perhaps be connected also with Aristotle's statement about dividing figures in *Met.* IX, 9. But clearly this was not, for the majority of late writers, the standard 'analysis', even when analysis was understood as a path from the complex to the simple. More often the path from complex to the simple parts in the geometrical analysis was interpreted as a path from complex *propositions* to their simple parts. Very often these simple 'parts' were interpreted as propositions and they tended to get mixed with 'simple' (self-evident) propositions, that is, with axioms.

In a passage of Proclus (Friedlein p. 244ff.) we are told that *theorems* are complex or simple. However, Proclus adds that a theorem is complex if and only if its hypotheses (the *dedomena*) or else its 'conclusion' (the *zetoumenon*) are complex. (Interestingly, Proclus says that the geometers favor complex propositions because they are easier to analyse, thus apparently betraying his awareness of the need of auxiliary constructions in analysis. See p. 246.) This interpretation is related to our analysis of figures. For it is the internal structure of the proposition in question that is being analysed here, that is, what the proposition says about the details of the configuration it speaks of. This analysis also ended ultimately at the 'principles' (*arkhai*: primitive terms? primitive propositions?; cf. *Proclus*, ed. by Friedlein, p. 246, lines 8–12).

But often the internal structure of a proposition (to be analysed) does not figure in the description of the method. Thus analysis becomes simply a passage from a complex proposition (which we want to analyse) to the axioms (simple propositions) or to previously established theorems. Sometimes only a statement to this effect is given but nothing is said about the crucial question as to how these elements of analyses are linked together. When something *is* said about the linkage, we sometimes seem to have something like our full-ledged propositional interpretation outlined on p. 32, occasionally expressed (as far as possible) in syllogistic terms.[42]

A passage in Proclus (this time the name of Porphyry – probably P. the Neoplatonist – is mentioned) is clearer in this respect than most others.[43] Proclus is speaking of *reductio ad absurdum* arguments, and of analysis. The reductive argument destroys its hypothesis, and thus "confirms the proposition in the beginning" (τὸ ἐξ ἀρχῆς ζητούμενον, Friedlein p. 255, lines 10–12; here what is 'in the beginning' is the proposition we want to establish, say *P*).

But when Proclus proceeds and speaks also about analysis, it is seen that what is called an *arkhe* is now the counter-assumption, that is, in a *reductio* the *arkhe* is $\sim P$. When Proclus begins to explain the role of axioms in the *reductio ad absurdum* arguments and in analysis, the passage becomes a little confused. For the axioms are also called *arkhai* elsewhere in Proclus. However, so much can be said that Proclus thinks that the reductive argument uses a specific mode of inference, "the second type of hypothetical syllogism".[44] A closer inspection of the example Proclus gives shows that the 'syllogism' probably is (an instance of) the second undemonstrable argument of the propositional logic of the Stoics. Schematically:

$$(7) \qquad \frac{\begin{array}{l} \sim P \supset \sim Q \\ \sim \sim Q \end{array}}{P} \qquad (= Q)$$

Here P is again the proposition we want to prove, and Q is an axiom (in the example of Proclus, "the whole is not equal to the parts").

If this way of understanding Proclus is correct in the main, he in fact succeeds in reconciling the method of analysis with the idea of 'downward' movement and in connecting it with *reductio ad absurdum*, but only at the expense of interpreting the startingpoint of an analysis as the *negation* of the desired theorem.

Proclus' passage is rather confused, however. Two different ideas (the one borrowed ultimately from the Aristotelian theory of science, and the other, a Stoic one) seem to be mixed with each other in Proclus. In any case Proclus is presupposing that there is a 'syllogism' in an analysis, too. (Cf. Friedlein, p. 256, line 1.) That is, he is presupposing that analysis can be interpreted in terms of propositional logic. He does not say, however, what mode of argument is used in an analysis.

The ideas we have discussed (analysis as an analysis of figures, analysis as an analysis of the complex 'enunciation' of a proposition, and analysis as a propositional inference or as an inference of syllogistical logic) are all (taken separately) justifiable to a certain extent, and (taken together) not wholly incompatible. That several of them are present in one and the same author – Proclus – is not as such very misleading. But Proclus' grasp of the geometrical method is nevertheless far from satisfactory, for reasons which we shall try to spell out.

Much of what can be said about Proclus applies also to several other philosophical writers we have discussed above. For one thing, the late writers were not very fond of sharp classifications (compared, e.g., with the old Stoics). Partly this may be an indication of underdeveloped technical terminology. As we have seen, for instance the concept of *arkhe*, the end-point of analysis, is in Proclus (and often elsewhere) rather fluid. But often it seems that the different elements in mathematics are simply not distinguished from each other by authors such as Proclus. (For instance, when Proclus speaks of the *arkhai* of mathematics, it is not clear whether mathematical axioms, postulates, and definitions are distinguished by him from figures, magnitudes, and properties.[45]) Owing to this genuine lack of category distinctions in discussing mathematics, the nature of the different ingredients of analysis is not spelled out clearly by Proclus.

There were other difficulties. The standard logical tools of ancient philosophers could never really cope with the actual mathematical practice. Neither the propositional calculus of the Stoics nor the Aristotelian syllogistic (which after all is no more powerful than monadic first-order predicate calculus), could express, even in principle, the richness of the mathematics of the day. But this is precisely what many of the philosophers and commentators we have discussed tried to do. In particular, the use of *ekthesis* (or instantiation), which is the hallmark of all ancient geometrical practice, analytical or synthetical, had no place in propositional logic and was never fully incorporated into the Aristotelian syllogistic or clearly understood by Peripatetic logicians. An inevitable consequence was that essential aspects of the conceptual situation were lost sight of. An average philosopher *qua* philosopher simply could not describe the richness of the ideas on which geometrical analysis was based.

Another point is less obvious, and it is important for our appreciation of the evidence we find in Proclus. The basic doctrine of the Neoplatonists was a hierarchy of beings. This easily led them not just to overlook such things as constructions and problems but to assign to constructions little value in their philosophy of mathematics. (Of course a construction, being a *genesis*, could not rank very high in the Neoplatonistic hierarchy of beings.) This is the way in which, e.g., Proclus views the situation, in spite of his professed impartiality. But this downgrading of constructions is misleading (cf. Ch. I above), and it led Proclus to other inaccuracies. His (or his source's, Carpus') remark to the effect that problems are easier than theorems is mistaken historically and probably even theoretically.[46] For the three most famous geometrical questions in the Antiquity (the duplication of the cube, the trisection of the angle, and the squaring the circle) were surely problems and very 'difficult' problems indeed. Theoretically, if problems are thought of as existential propositions, they apparently are on plausible criteria at least as difficult as theorems (universal propositions).[47] (The situation is not altered if we adopt a 'constructional' standpoint and prefer to think in terms of processes of construction instead of existential propositions. Presumably something like this view is what Proclus has in mind in the passage treated above.)

This philosophically motivated suspicion of constructions (conspicuously exemplified by Proclus' neglect of problems) of course affects also descriptions of analysis, especially in conjunction with the limited power of the available logical tools. Proclus does not avoid subjects like constructions. But he, like other philosophers of the same ilk, is not greatly interested in such less 'valuable' topics, and consequently does not present a satisfactory picture of them. Incidentally he does not even mention Euclid's *Data* in his catalogue of Euclid's works in his commentary on the first book of the *Elements*.[48] Small wonder, then, that auxiliary constructions and the 'resolution' which deals with them (typically using theorems from the *Data*) have no place in his comments on the method of analysis. And yet Proclus quotes approvingly the opinion of Carpus which definitely connects the use of the analytical method with problems (see above).[49]

Pappus, in treating theoretical and problematical analysis on a par, is clearly not committed to the fallacy which we in its most conspicuous form find in Proclus. Similarly, there are in his description of the method of analysis and synthesis no traces of the reliance on the restrictive theoretical models we have discussed. In spite of this, we cannot wholly attribute Pappus' description to his experience as a practising mathematician. For as we have seen, there *are* philosophical (or methodological) ingredients in his description of analysis and synthesis.

This observation is interesting and important. But what positive can be said about the background of these ingredients, over and above the fact that the restrictive models are not present in them? Strictly speaking, the attribution of Peripatetic and Platonic influences to Pappus is almost vacuous. In Pappus' time, and in the immediately preceding centuries, influences of this sort were commonplace. Knowing the distant origin of such commonplace ideas is not very helpful. We would like to know what shape was given to the old ideas by Pappus' immediate sources and by Pappus himself, respectively, at least in general outline. Part of the matter can be explained by pointing out that Pappus' description is a kind of preface to a technical book (or perhaps to a chapter of a book). In prefaces of this kind (they are rather common in ancient mathematical literature) well-known methodological or philosophical ideas had always been resorted to, whereas technical details were omitted for obvious expositional reasons.

These prefaces and introductions are almost without exception programmatic in one way or another. In this study, we have tried to show that Pappus' description does not entail a methodological allegiance to any particular philosophical school, be it Stoic, Neoplatonic, or Peripatetic. Pappus says (Hultsch 634, line 6) that analysis is a heuristical method ἐν γραμμαῖς, that is, in geometrical investigations (literally 'in the study of lines'). A mere graphical study is not what is meant, but a serious theoretical investigation, geometrical in nature. For instance Ptolemy (whom Pappus commented on) speaks of proof 'by means of lines' in this sense.[50] Pappus (not unexpectedly) sees in the geometrical analysis a geometrical method, and not, say, an instance of a general philosophical method. His description as a whole bears out this explanation.

In this context it is useful to notice that geometrical proofs were sometimes contrasted to other kinds of proofs. Thus Galen says frankly that he had found the Stoic, Peripatetic, and Platonic – he knew Albinus personally – logical theories impractical and useless to many purposes. What he recommends as a substitute is precisely the *mos geometricus*, geometrical or 'linear' proof (γραμμικὴ ἀπόδειξις). Galen wanted to make the geometrical way of proving theorems a part of a general methodological theory, and studied to a certain extent its peculiarities (notably the role of axioms in a mathematical theory, and relational arguments).[51] As we have seen, it appears that he even contrasted the geometrical analysis to the Stoic propositional 'analysis' (see above).

Galen says that he learned the value of 'linear' proofs from the teaching of some relatives who were mathematicians.[52] In those parts of his own logical studies which fall outside the conventional logical theory, he also refers to earlier studies by Posidonius. But clearly there must have been still other studies.[53] (Of course, some attention was paid to the peculiarities of mathematical reasoning already by Plato, by Aristotle, and by mathematicians like Menaechmus, as we have seen. But the emphasis on geometrical proofs as distinguished from other kinds of proof which we find in Galen seems to be a much later phenomenon.)

In spite of Galen's emphasis on geometrical proofs, the earlier methodological and logical traditions were nevertheless extensively used by him even in his 'linear' studies. In this respect Pappus' description of analysis and synthesis is to a certain extent similar to Galen.

Of course, we cannot possibly hope to locate the immediate source of the different parts of Pappus' description, Pappus knew Heron very well and likewise knew Geminus' work 'On the Classification of Geometry'.[54] Pappus knew also Carpus' *mechanicus*. All these men had written on geometrical analysis. But nothing we know suggests that Pappus' description can be completely understood against this background. The history of geometry by Eudemus might have been the source of the Peripatetic elements in Pappus, but Eudemus is not mentioned in the *Collectio*. Furthermore, Pappus knew personally at least one philosopher of his time (Hultsch 35), and could have been acquainted with the original works of Aristotle and Plato as well as with later conceptions of philosophical analysis. Lastly, there is Pappus' own contribution. We must for instance bear in mind that Pappus was, as far as we know, the first person to discuss the part of analysis we have called 'resolution'.[56]

NOTES

[1] By 'tradition' we do not mean an uninterrupted, unerring carrying-over of information from one to another writer. 'Tradition' here is not at all immune to changes of the original content.

[2] Cf. *Proclus* (ed. Friedlein), p. 211, line 18ff.; Diogenes Laertius II, 24; *Academicorum Philosophorum Index Herculanensis*, ed. S. Mekler, Weidmann, Berlin, 1902, p. 17 (= Col. Y, 14ff.).

[3] See T. L. Heath, *Greek Mathematics*, Clarendon Press, Oxford, 1921, Vol. I, p. 291.

[4] *Archimedes* (ed. Heiberg), Vol. III, p. 92ff. – Concerning the evidence of the early Pythagoreans and of Hippocrates of Chios, see Holger Thesleff, 'Scientific and Technical Style in Early Greek Prose', *Arctos* (N.S.) 8 (1966), 89–113, see p. 105; and for the material offered by the Platonic dialogues, see W. R. Knorr, *The Pre-Euclidean Theory of Incommensurable Magnitudes*, diss., Harvard, 1972, Section VI.

[5] However, the Greek names of conic sections used are not genuine. See Heath, *Greek Mathematics*, Vol. I, p. 255. – These analytical arguments are similar to the problematical ones of Pappus.

6 For the pentagon, see Heath, *Greek Mathematics*, Vol. I, p. 168; for the duplication of the cube, p. 183.

7 Friedlein, p. 242, line 14ff. The relative dates of Carpus and Heron are not clear. See Fr. Krafft, *Dynamische und statistische Betrachtungsweise in der antiken Mechanik*, Franz Steiner Verlag GMBH, Wiesbaden, 1970, p. 100, and Heath, *Elements*, Vol. I, p. 21.

8 It has been suggested (first by Bretschneider) that the method of the theoretical analyses interpolated into Euclid (see Ch. III, note 4 above) is due to Eudoxus. See Heath, *Elements*, Vol. III, p. 442; E. Sachs, *Die fünf platonischen Körper*, Weidmann, Berlin, 1917 (Philologische Untersuchungen, 24), p. 97; F. Lasserre, *Die Fragmente des Eudoxos von Knidos*, Walter De Gruyter & Co., Berlin (Texte und Kommentare, 4), p. 177 and 37. In our opinion, however, it is safer to connect these analyses with the similar ones *apud* an-Nairīzī which we have than with the hypothetical theoretical analyses of Eudoxus. Notice, however, that this policy does not just deny the value of the suggestion of Bretschneider. But clearly further study is needed here – and much caution.

9 See Heath, *Greek Mathematics*, Vol. I, p. 291.

10 See H. G. Zeuthen, *Forelaesning over Mathematikens Historie. Oldtiden*, Høst & Søns Forlag, Copenhagen, 1949, p. 96.

11 See, e.g., Hultsch 918, line 11; 278, 18; 288, 12, and *Heron* (ed. Schöne), Vol. III, p. 170, line 2; III, p. 166, line 13.

In these examples, the reduction (*apagoge*) is not an inference from the desired proposition (say *P*) to a known proposition (say *Q*). In these examples, after an analysis proper one proceeds in the 'resolution', starting from the originally given entities, and shows that other entities are given, and happens to find out (by help of the functional dependencies which were detected in the analysis proper) that a known proposition *Q* will be useful in the synthesis as a premise. In the synthesis, then, *Q* in fact will be a premise (among others).

12 See *Meteor*. III, 5, 376a3ff. There we see that Aristotle already knew the Props. 1, 25, and 26 of the *Data*. Cf. J. L. Heiberg, 'Mathematisches zu Aristoteles', *Abhandlungen zur Geschichte der mathematischen Wissenschaften* 18 (1904), 3–49; esp. 27–28. – The authenticity of Book III of the *Meteorologica* has not been seriously contested. Heiberg thinks that Euclid's *Data* was the first collection of its type, but that the material of the book was created earlier. See *Litteraturgeschichtliche Studien über Euklid*, Teubner, Leipzig, 1882, p. 40.

13 See Heath, *Greek Mathematics*, Vol. II, p. 358–360.

14 Aristotle's treatment of the negative outcome is perhaps *prima facie* not completely clear. For he appears not to say that a negative outcome (an impossibility arising in the course of analysis) shows that the desired result is impossible. He says only that "if we come on an impossibility, we give up the search" (1112b24–25).

However, this is enough to show that Aristotle's description of the negative outcome is analogous with Pappus'. For if only one of the several possible ways of realizing the desired *telos* turns out to be impossible, there is no good reason to give up the search, unless such an impossibility somehow suffices to show the impossibility of the *telos*. (What Aristotle ought to have said is that if we come upon an impossibility *in every direction*, we give up the search.)

In this confusion of Aristotle's we probably have an instance of the pernicious effects of the directional or propositional interpretation. If the distinction between analysis and synthesis lies in the direction of the procedure, little interest is aroused by the possibility of there being several parallel courses that an analysis might take.

In Heath's posthumous book, *Mathematics in Aristotle*, Clarendon Press, Oxford, 1949, pp. 270–271, the problem of the different outcomes of analysis is not treated.

¹⁵ Aristotle's passage (and related Aristotelian material) and its philosophical implications have been dealt with by Jaakko Hintikka, 'Practical vs. Theoretical Reason: An Ambiguous Legacy', in Hintikka, *Knowledge and the Known: Historical Perspectives in Epistemology*, D. Reidel, Dordrecht and Boston, 1974 (*Synthese Historical Library*, Vol. 11), pp. 80–97.

¹⁶ Aristotle's description of the possible outcomes of *bouleusis* is perhaps more carefully formulated than that in Pappus' description about the outcomes of analysis (see *Eth. Nic.* III. 3, 1112b24–27), but not entirely flawless. Cf. p. 78 of our study and note 14 above.

¹⁷ Cf. *In Phys.* (CAG XVI), p. 335, line 16.

¹⁸ Cf. *In An. Post.* I, p. 121, line 8ff.

¹⁹ Cf. *In An. Post.* I, p. 120, line 20.

²⁰ *In Phys.* p. 335, line 32.

²¹ Cf. Hultsch 634, lines 11–13 and 636, lines 1–4; 7–10.

²² Cf. Heiberg *Euclid* Vol. IV, p. 364, line 17ff.

²³ Cf. Heath, *Elements*, Vol. III, p. 442.

²⁴ Cf. Albinus' *Epitomé* (ed. Pierre Louis), Imprimeries Réunies, Rennes, 1945, p. 27. – For Albinus and his background, see R. E. Witt, *Albinus and the History of Middle Platonism*, Cambridge at the University Press, 1937.

²⁵ So also Gulley, *op. cit.* (in Ch. II, note 4), p. 6, note 1.

²⁶ *Epitomé* p. 35. ἡγούμενον and λῆγον here are the Stoic terms for antecedent and consequent (in an implication). See Mates *op. cit.* (in Ch. II, note 8), 'Glossary'.

²⁷ Cf. *Phaedo* 100Aff.; *Rep.* VI, 510Bff. For further references, see, e.g., S. Scolnicov, *Plato's Method of Hypothesis in the Middle Dialogues*, unpublished Ph. D. thesis, Cambridge University, 1973.

²⁸ We recall that in the *Republic* VI, 510Bff. the use of hypotheses is connected with geometrical investigations. Cf. also the evidence mentioned in note 2 above, and note 8.

²⁹ See N. Gilbert, *Renaissance Concepts of Method*, p. 34, note 52. Gilbert finds it tempting to consider "the possibility of some infiltration of Stoic logic into this geometrical methodology [of Pappus]".

³⁰ See Mates, *op. cit.*, p. 77, and Oskar Becker, 'Zwei Untersuchungen zur antiken Logik', *Klassisch-Philol. Studien*, Vol. 17, Otto Harrassowitz, Wiesbaden, 1957, pp. 27–49.

³¹ See *Medicorum Graecorum Opera* (ed. Kühn), Vol. XIX, p. 47, line 15. Galen knew the Stoic 'analysis' in the sense presupposed here. See Becker, *op. cit.* above, p. 36. It is of course conceivable that the geometrical 'analysis' (*analytike*) in the title of Galen's treatise does not refer to geometrical analysis proper directly but to geometrical 'analytics' in the wider sense, i.e., to geometrical proofs in general.

³² See Gilbert, *Renaissance Concepts of Method*, p. 20 and p. 32ff. (especially for Galen's 'analytical' methodology; the relations of this methodology and of other 'analyses' known to Galen pose several problems), and Gulley, 'Greek Geometrical Analysis'. Thus Albinus knew several kinds of philosophical analysis (see *Epitomé* p. 25), and so did others. See, e.g., Ammonius *In An. Pr.* I (CAG IV, 6), p. 5, line 5ff.; Alexander *In An. Pr.* I (CAG II, 1). p. 7; Philoponus *In An. Post.* II (CAG XIII, 3), pp. 335–336. Even Clement of Alexandria speaks of analysis (Migne Vol. IX, p. 568, line 27ff.)

[33] Geminus *apud* Ammonius (CAG IV, 6) *In An. Pr.* I, p. 5, line 28. See Max P. C. Schmidt, 'Philologische Beiträge zu griechischen Mathematikern', *Philologus* **42** (1884), 82–118 for the Stoic Geminus.

[34] See *Anaritius* (ed. Curtze) p. 88ff., and *Codex Leidensis*, 399, 1 (ed. Besthorn and Heiberg), Vol. II, 1, p. 7ff. See also Ammonius *In An. Pr.* I (CAG IV, 6), p. 5, lines 26–27 where these analyses are probably referred to (without mentioning Heron), and Heiberg *Euclid* Vol. V, pp. 230–231 where an analysis (by Heron?) of *Elements* 2, 3 is (partly) preserved. – See Heath, *Elements*, Vol. I, p. 373 for the quotation.

[35] See *Anaritius* (Curtze), p. 89, and *Codex Leidensis* 399, 1, p. 9. See also Heiberg *Euclid* Vol. V, p. 675 where a scholion with partly similar content can be found.

[36] See Ch. III, note 4 on Heron.

[37] Gherard's translation (*Anaritius*): "… reducemus ad illam, cuius probatio iam precessit". Besthorn-Heiberg: "… in rem iam antea demonstratam dissoluimus". The scholion (*Euclid*, Vol. V, 675) reads: … ἐπί τι γνώριμον τῶν ἤδη προαποδεδειγμένων.

[38] *Heron* (ed. Schöne), Vol. III, p. 30, line 30, and p. 152, line 19.

[39] See for *arkhe* Heron (ed. Schöne), Vol. III, p. 114, line 17 and 27; Vol. III, p. 158, line 18.

[40] *In An. Pr.* (CAG XIII, 2), p. 5, lines 16–17.

[41] Friedlein, p. 118, line 15; p. 442, lines 13 and 21.

[42] Cf., e.g., an anonymous commentator on *Soph. El.* (CAG XXIII, 4), p. 41, line 34 and p. 47, line 21. Probably also *Proclus* (ed. Friedlein), p. 8, lines 5–8 belongs to this group.

[43] The fluid meaning of the term *arkhe* comes nicely out at *Proclus* (ed. Friedlein), p. 255, line 12ff. – Here we also see that Proclus does not completely succeed in reconciling his different sources. (For Proclus' sources, see Heath, *Elements*, Vol. I, pp. 29–45.) This is not unusual in Proclus.

[44] Friedlein, p. 256, lines 3–6.

[45] Friedlein, p. 57.

[46] Friedlein, p. 242.

[47] In terms of decidability, it might even be that in some geometrical systems the theorems (conceived of as universal propositions) are less 'difficult'. Cf. A. Tarski, 'What Is Elementary Geometry?', in *The Axiomatic Method* (see Ch. I, note 16), p. 27, and on the sense of 'universal', here p. 24.

Notice that even if Tarski's conjecture (*op. cit.*, p. 27, lines 17–20) proves to be correct, this does not contradict what we said earlier (p. 64ff.). The greater difficulty of theorems as compared with problems concerns the proof, not an intrinsic difficulty of answering the two kinds of questions.

[48] See Friedlein, p. 69. See also Heiberg, *Litteraturgeschichtliche Studien über Euklid*, p. 173. – Marinus the commentator of the *Data* was Proclus' pupil, however. See Morrow, *op. cit.* p. xxii. But Marinus refers in his commentary not to Proclus' views but to Pappus.

[49] Cf. p. 242 (Friedlein), lines 14–17.

[50] See I. Thomas, *Greek Mathematical Works*, Vol. II, pp. 412–414. For Pappus' commentary on Ptolemy, see Ch. VI, note 12.

[51] Cf. Kühn, Vol. XIX, p. 39ff., and I. von Müller, 'Ueber Galens Werk vom Wissenschaftlichen Beweis', *Abh. der Bayer. Akad. der Wissenschaften*, Philos.-Philol. Kl., Vol. 20 (1897), pp. 403–478. For Galen's treatment of relations, see J. Mau, *Galen: Einführung in die Logik. Kommentar.*, Akademie-Verlag, Berlin, 1960, p. 52ff. – We have seen that the 'ekthetical' nature of geometrical proofs and the role of constructions

in them had been appreciated, e.g., by certain sources of Proclus'. There is a Galenian *ekthesis* (cf. E. Beth, *Aspects of Modern Logic*, p. 138) but it seems to be purely syllogistical in nature, as we have it, and not related to 'linear' proofs.

[52] See Kühn, Vol. XIX, p. 40. In spite of Galen's outspoken predilection of geometrical proofs, it seems that his 'analytical' method (cf. note 32 above) which is said to proceed "from the notion of the goal (*telos*)" was not modelled solely after the geometrical method. However, it *was* interpreted as the same as the method of geometers already by the eleventh century Arab, Ali ibn Ridwan, who was influential also in the medieval West. Cf. William F. Edwards, 'Scientific Method in the School of Padua', in *Naturalism and Historical Understanding*, John P. Anton (ed.), State University of New York Press, Albany, N.Y., 1967, pp. 53–68. – It is interesting that ibn Ridwan tried also to assimilate the analysis/synthesis distinction (of Galen and of geometers') to the distinction of *hoti/dioti* demonstration in *An. Post.* of the Stagirite.

[53] Cf. *Institutio Logica* (ed. C. Kalbfleisch), Teubner, Leipzig, 1896, p. 47.

[54] Hultsch, 1026, line 8ff.

[55] Pappus does not mention Eudemus; cf. Konrat Ziegler's 'Pappos', article in Pauly-Wissowa, and Heath, *Greek Mathematics*, Vol. II, p. 358ff. for Pappus' sources.

[56] Cf. *Data* (ed. Menge), p. 256.

CHAPTER IX

ON THE SIGNIFICANCE OF THE METHOD OF
ANALYSIS IN EARLY MODERN SCIENCE

It has been repeatedly suggested, among others by Ernst Cassirer,[1] J. H. Randall, Jr.,[2] and Oskar Becker,[3] that the method of analysis and synthesis constitutes the methodological 'secret' of the leading scientists of the heroic period from Galileo to Newton. Here we shall not argue for or against this sweeping thesis, but rather try to put it into a perspective and to offer some indirect support to it by examining (i) how the method in question is related to the geometrical analysis of the Greeks and (ii) how understanding the nature of this analysis throws light on certain problems concerning early modern science and early modern philosophy.

Many important aspects of the methodological ideas of Galileo, Descartes, and Newton can in fact be related to what we have found about the Greek geometrical analysis. It is not entirely surprising that there should be connections here. Newton is known to have been a great admirer of Greek geometry,[4] and although this predilection is often interpreted as a preference of synthetic over analytic methods, Newton's admiration certainly seems to have encompassed Greek analysis as well.[5] Newton's *Arithmetica universalis* "contains many extensions of the Pappan *Treasury of Analysis*, together with much praise of ancient methods."[6] Descartes, in his *La Géométrie*, likewise takes off from problems posed by Pappus.[7] However, even in view of such obvious external connections, the extent of the influence of the analytical paradigm is truly remarkable. The most important, and probably best known, methodological statement by Newton shows amply this influence.[8]

As in Mathematicks, so in Natural Philosophy, the Investigation of difficult Things by the Method of Analysis, ought ever to precede the Method of Composition. This Analysis consists in making Experiments and Observations, and in drawing general Conclusions from them by Induction, and admitting of no Objections against the Conclusions, but such as are taken from Experiments, or other certain Truths. For Hypotheses are not to be regarded in experimental Philosophy. And although the arguing from Experiments and Observations by Induction be no Demonstration of general Conclusions; yet it is the best way of arguing which the Nature of Things admits of, and may be looked upon as so much the stronger, by how much the Induction is more general. And if no Exceptions occur from Phaenomena, the Conclusion may be pronounced generally. But if at any time afterwards any Exception shall occur from Experiments, it may then begin to be pronounced with such Exceptions as occur. By this way of Analysis we may proceed from Compounds to Ingredients, and from Motions to the Forces producing them; and in general, from Effects to their Causes, and from particular Causes to more general ones, till the Argument end in the most general. This is the Method of Analysis: And the Synthesis consists in assuming the Causes discover'd, and establish'd as Principles, and by them explaining the Phaenomena proceeding from them, and proving the Explanations.

Here Newton goes as far as to identify in effect his own experimental method with the analytical method of geometers. The coherence and force of the methodological views expressed by Newton in the quoted passage are often underestimated in our days, it seems to us. It is not always understood what the common denominator of the two methods mentioned by Newton, the mathematical analysis and the experimental one, is supposed by him to be. However, from the vantage point we have already reached, the connection is obvious.[9] Newton, like any experienced mathematician, is thinking of the geometrical analysis as an analysis of figures, that is to say, as a systematical study of the interdependencies of the geometrical objects in a given configuration, including both the 'known' (controllable) and 'unknown' (uncontrollable) factors. From this idea it is but a short step to conceiving of the analytical procedure as a general method of studying such 'dynamical' interdependencies, making no difference between the known and unknown elements. It was in this sense that the concept of analysis was extended to other parts of mathematics. For instance, algebra was considered analytical because in it one studied the interdependencies between a number of quantities, some of them unknown, by means of equations.[10] From this, it was but a short step to the idea that an experimental setup represented a kind of analytical situation, too, in that what is happening in a typical controlled experiment is a study of what depends on what in it – and hopefully also precisely what mathematical relationships these dependencies exemplify. Newton's conception of the experimental method as a kind of analysis is thus an outgrowth of the idea of analysis as an analysis of figures or more generally of geometrical configurations. Newton was trying to analyse an experimental situation in the same way as a Greek geometer like Pappus was trying to analyse a figure in the sense of trying to establish the interrelations of its several parts.

All this is a far cry from the idea of analysis as an analysis of deductive connections.[11]

In Newton's statement, we can catch a glimpse of the terminology in which the different ideas of analysis were typically expressed in Newton's time. The 'analysis of configurations' idea was expressed for instance by speaking of proceeding 'from Compounds to Ingredients' whereas the old 'directional' idea or the idea of analysis as an analysis of propositional connections rears its head when Newton speaks of proceeding 'from Effects to their Causes'. (This relationship was in the old Aristotelian thinking assimilated closely to the relationship of consequences to their premisses.) However, Newton's 'Causes' are not any more of the nature of antecedent conditions capable of acting as middle terms in Aristotelian syllogisms. They, too, are physical factors influencing the concrete situation in question, for instance (*apud* Newton's own statement) "Forces producing [Motions]". Hence even the cause-effect terminology does not mark a real reversal to the propositional and directional interpretation of analysis in Newton. (This relation of the cause-effect order to the order of analysis also marks one of sharpest differences between Newton and Hooke; see note 11 above.)

Several important observations can be made here on the methodological role of analysis in Newton and his predecessors.

(1) It seems to us that a shift from the propositional (directional) interpretation of analysis to the constructional (analysis-of-configurations) interpretation went hand in hand with a revival of analysis as an important methodological tool. Thus the former development could provide important indicators of what precise significance we should attach to sundry apparent anticipations of the Galilean method of resolution and composition (analysis and synthesis) in the earlier literature.[12] Many of those uses of the *resolutio-compositio* terminology which have actually been pointed out in the literature still remained firmly within the purview of the Aristotelian interpretation, that is, the directional and propositional interpretation, of analysis and synthesis, which seems to have dominated the discussion in the Middle Ages.[13] Sometimes all connection with the geometrical and other mathematical senses of resolution (analysis) was explicitly disclaimed in the medieval literature.[14] In view of the close relationship of Galileo's and Newton's ideas of the role of analysis in experimental science to the ancient geometrical analysis, such apparent

anticipations of Galileo's and Newton's methodology cannot be considered very significant. As Neal Gilbert points out, a case in point is Zabarella's vaunted anticipation of the Galilean method of resolution and composition: Zabarella "expressly excludes mathematical analysis from his resolutive method...", Gilbert notes.[15] Professor Gilbert also points out in effect that the typical terminology of analysis-of-configurations, such as 'analysing a phenomenon into its components', does not represent accurately Zabarella's jargon.

The same goes for Grosseteste's use of the language of resolution and composition. He, too, contrasts in so many words the procedure of the natural sciences from 'what is more knowable to us' to 'what is more knowable in itself' to the procedure of mathematicians in whose field the two coincide – thus leaving in Grosseteste's opinion no room for analysis (*resolutio*) in mathematics, it seems.[16]

(2) It has not always been appreciated precisely where the similarities and dissimililarities between Newton and Galileo lie. For instance, Henry Guerlac suggests that for Galileo, unlike Newton, "the analysis by experiment and observation is merely suggestive or indicative".[17] This is scarcely borne out by what Galileo actually says – or, rather, lets his characters say:[18]

Simplicio: Aristotle first laid the basis of his argument *a priori*, showing the necessity of the inalterability of heaven by means of natural, evident, and clear principles. He afterwards supported the same *a posteriori*, by the senses and by the traditions of the ancients.

Salviati: What you refer to is the method he uses in writing his doctrine, but I do not believe it to be that with which he investigated it. Rather, I think it certain that he first obtained it by means of the senses, experiments, and observations, to assure himself as much as possible of his conclusions. Afterwards he sought means to make them demonstrable. This is what is done for the most part in the demonstrative sciences; this comes about because when the conclusion is true, one may by making use of the analytical methods [*metodo risolutivo*] hit upon some proposition which is already demonstrated, or arrive at some axiomatic principle.... And you may be sure that Pythagoras, long before he discovered the proof for which he sacrificed a hecatomb, was sure that the square on the side opposite the right angle of a right triangle was equal to the squares on the other two sides. The certainty of a conclusion assists not a little in the discovery of its proof....

Galileo's description of the *metodo risolutivo* reminds us of Pappus' words. What is more important, the only qualifications he has concerning the conclusiveness of "senses, experiments and observations" are precisely parallell to Newton's:[19]

Galileo:

[Aristotle first obtained his doctrine] by means of senses, experiments, and observations, to assure himself *as much as possible* of his conclusions.

Newton:

And although the arguing from Experiments and Observations by Induction be no Demonstration of General Conclusions; yet it is *the best way of arguing* which the Nature of Things admits of...

If there is a difference between Newton and Galileo here, it is not in what they think analysis or resolution can accomplish, but in what they think of synthesis (composition). Galileo did not completely disentangle himself from the old Aristotelian idea that the last, indivisible steps uncovered by analysis can be turned around so as to become in the last instance *conceptual* truths, in fact definitions of sorts.[20] In other words, for him the inverted synthetic proof is still comparable to a geometrical one not only in form but only with respect to the nature of its several steps.[21] For Newton, synthesis meant simply putting together new, often more complex configurations by means of the general laws which had been uncovered in the analysis and generalized in the induction. The certainty of such synthesis was for him no greater and no less than that of these inductive generalizations.

(3) Only in the context of Newton's interpretation of the experimental method as analysis can we understand his famous strictures against hypotheses.[22] One from of it occurs already in our first quotation from him. ("For Hypotheses are not to be regarded in experimental Philosophy.") Another form is found in an unpublished MS which Henry Guerlac has quoted.[23] After having explained the method of analysis and composition along the same lines as in our passage, Newton writes: "But if without deriving the properties of things from Phaenomena you feign Hypotheses... your systeme will be little better than a Romance". It is clear that by hypotheses Newton meant, when he was condemning them, assumptions which are not generalized from the results of an analysis of the interrelations of the different factors of a suitable experimental setup or a comparable situation.[24] What was wrong with them, according to Newton, is not that they are not conceptual truths, but that they are not obtained analytically. Whatever hesitations and changes Newton's

views of hypotheses may have undergone, this is the basic and the focal point of his negative attitude to them.

(4) More generally, in terms of the analytical procedure we can perhaps put Newton's method in its place better than in standard accounts. This is not quite as simple a matter as might first seem. Newton's method differs from the usual descriptions of the so-called hypothetico-deductive method precisely in the way his strictures against hypotheses bring out. For him, not any hypothesis having testable deductive consequences is acceptable. He does not allow any old hypotheses in our sense of the word, but only those that have been inferred, derived, or, as he sometimes put it, 'deduced' from phenomena. He is not only concerned to provide a set of rules for confirming hypotheses and theories already known, but also to outline a rational method of finding these hypotheses and theories in the first place. Hence Robert Palter was not entirely wrong when he recently contrasted Newton's method to the hypothetico-deductive one.[25]

However, we cannot comfortably pidgeonhole Newton's (and Galileo's) analytical method in the box labelled 'inductive method', either, as Palter strives to do. Induction is but one of the steps in the Newtonian scheme, and a comparatively trivial one at that. The Newtonian method may perhaps be schematized thus:

(i) an analysis of a certain situation into its ingredients and factors →

(ii) an examination of the interdependencies between these factors →

(iii) a generalization of the relationships so discovered to all similar situations →

(iv) deductive applications of these general laws to explain and to predict other situations.

Here induction occurs only as step (iii). And an induction so conceived of seems to have little to do with any of the procedures studied by the theorists of induction. A step like (iii) cannot very well be thought of either as induction by enumeration or induction by elimination. In some of his methodological pronouncements Newton admittedly speaks of step (ii) as involving the examination of several experimental or observational situations. It is nevertheless clear that although such comparisons may be involved, they need not. In fact, Newton "insisted upon the cogency of a *single*, well-contrived experiment to answer a specific question, as opposed to the Baconian procedure of collecting and comparing innumerable 'instances' of a phenomenon" (Guerlac). It is not for nothing that Newton was instrumental in giving the phrase 'crucial experiment' the currency it has enjoyed ever since. Surely a good inductivist ought to eschew Newton and side rather with Hooke who in his controversy with Newton who tried to appeal to 'many hundreds of trials' against Newton's lone *experimentum crucis*.[26] If anything, Newton thus appears more anti-inductivist than inductivist, especially as the role of the crucial experiment apparently must have been for him, not the elimination of competing hypotheses, but the very formation of the law to be "rendered general by induction".

The fact is that Newton's method is not very easy to describe by means of the commonplaces of the contemporary philosophy of science. Especially stage (i), which we have separated from the rest of analysis, viz. from (ii), involves a conceptualizing element which is not easily discussed in the terms of ready-made languages presupposed in most applications of modern philosophy of science. The best tools for handling it seem to us to be the (so far rather rudimentary) logical studies of conceptual enrichment and theory change.

It is to be noted that for Newton induction meant just the generalizing step. Experimental conclusions are "rendered general by induction", he says. For Newton, induction is not, Professor Mandelbaum's formulation notwithstanding,[27] "application of his theory to other cases to test its adequacy". This is the role he ascribes to synthesis.

It has been suggested that Newton was an inductivist because he believed that experimentation and observation can make laws highly certain – if not completely so. The trouble with this point is that before the inductive step there is no general law to be certain about. And after the inductive generalization we have nothing more certain in our hands than generalization itself, which cannot convey to 'synthetic' conclusions drawn from it a higher certainty than it possesses itself.

(5) This does not mean, however, that modern conceptualizations and results cannot be used to throw light on the methods and methodological ideas of early modern scientists. On the contrary, the nature of the experimental method as a kind of partial analogue to the geometrical analysis of the Greeks helps to put their methodological problems into perspective in at least one important respect.

The experimental method was clearly thought of by Newton as a method of discovery of sorts. This idea is of course even more pronounced in Descartes, whose whole philosophical and scientific method can be thought of as a kind of generalization from his analytical method in geometry.[28] It is also clear that Descartes for one fancied – or at least hoped – that his method would be an effective one.[29] To what extent were such hopes justified? If they were not, precisely what went wrong?

Here the perspective outlined in Chapter I above serves us well. It was pointed out there that in a proof of a conclusion from certain premises we often have to consider more individuals in their relation to each other than either in the conclusion or in the premises. This is a fairly general feature of the logical situation, obtaining already in first-order logic. Moreover, the number of additional individuals needed is often recursively unpredictable.

In elementary geometry, the introduction of these auxiliary individuals is what *auxiliary constructions* mean. In sufficiently elementary parts of geometry, their number is predictable. However, this is not the case in general.

When the study of physical configurations is thought of along the same lines as an analysis of a geometrical figure, the same problem arises. Even if the general laws governing the situation are known, it may still be the case that they serve to account for certain aspects of the interaction only if enough ingredients of the configuration are taken into consideration. Moreover, there need not be any way of telling whether enough factors have already been brought in. If so, a generalized analytical method will not be an effective discovery produce, however useful it may be heuristically.

There is no mystery why this should be the case. It is simply the dilemma which confronts us already in first-order logic. It is also independent of all practical difficulties of controlling an experimental setup or formulating the general laws which govern it.

This shows clearly the theoretical limitions of the 'analytical' methodology of the first great modern scientists. However powerful a tool it may have been heuristically (for the same reasons as were discussed in Chapter IV above for the fruitfulness of the analytical method of ancient geometers), it was not and could not have been a foolproof discovery procedure in all circumstances.

It is of interest to see that the failure of such alleged discovery procedures in empirical sciences is closely related to certain basic reasons for the non-triviality of logical and mathematical reasoning. They, too, can be traced back to the need of auxiliary individuals needed in our arguments.

It is amusing to see how nonchalantly this crucial point is treated in Descartes' *La Géométrie*. The advice to problem-solvers given there is as follows:

"If, then, we wish to solve any problem, we first suppose the solution already effected, and *give names to all the lines that seem needful for its construction* – to those that are unknown as well as to those that are known." (Our italics; p. 299 of the original.)

There is nothing here that even resembles an effective discovery procedure.

In experimental research the general methodological problems just indicated sometimes come to the surface in the form of questions whether one's experimental setup really excludes 'disturbing factors' or whether it can be considered a 'closed system' for the purposes of experimentation.[30] It is important to realize that these are not just practical problems of marginal importance to a philosophical methodologist, but touch the very prospects of all systematic 'logic of scientific discovery'.

The general problem may perhaps be illustrated by recalling the familiar strategy of inferring the presence of a new, so far neglected factor from observed discrepancies with well-established laws. The discoveries of the outer planets have been the most spectacular, although not by any means the only, clear-cut cases in point. However, when the unknown factors are not observable, there in principle always remains the question whether apparent discrepancies between a theory and observations are due to presence of such an unanticipated ingredient or to a failure of the theory itself. (The early history of the study of the advance of the perihelion of Mercury may perhaps offer a partial illustration of this point.)

In Descartes' methodology, the problem is connected with his tacit assumption that the process of analysis he describes always comes to an end.

(6) The relationship of Descartes' scientific and philosophical method to the tradition of analysis is too intricate a subject to be dealt with in any detail here. It nevertheless seems to as that it can be used to elucidate Descartes' argumentation to a much larger extent than in the literature. A few remarks may help to appreciate Descartes' peculiarities.

(i) Descartes insists on discussing methodological matters in propositional terms or at least in terms of sequences of steps of thought. This has made it unnecessarily difficult to recognize applications of the analytical method in his philosophical argumentation.

(ii) In particular, although Descartes in so many words acknowledges that his *Meditations* were carried out analytically,[31] most discussions of the structure of the argumentation of this work are couched in deductive terms or in equivalent ones, for instance in terms of the justification of Descartes' belief in this or that (e.g., in mathematical truths, in God's existence, etc.). This is too narrow a framework to do justice to an analytically conducted argument.

For instance, the famous problem of the alleged 'Cartesian Circle' is little more than a special case of the general problem of turning an analytical argument around so as to make it into a strictly deductive one. Yet it has not recently been as much as connected with the problem of analysis and synthesis.[32]

(iii) A special case of the general analytical method is what might be called the method of limiting cases. Often, the study of the interrelations of a given geometrical or physical configuration can be highlighted – and especially important consequences drawn from it – by pushing one of the relevant factors to the limit, to infinity or to zero, as the case may be, more generally to its maximal or minimal value.

The significance of this strategy is well known, and an abundance of illustrations can be found for it in the history of science. Galileo's partial anticipation of the law of inertia is a case in point.

Certain features of Descartes' argumentation can be understood as applications of this method of limiting cases. The most important of them is of course his methodological doubt which is not a counsel of caution but an active attempt to push disbelief to the limit.

(7) Newton's relationship to the geometrical analysis of the Greeks deserves a special mention of its own. It seems to us that G. L. Huxley's emphasis is not quite appropriate when he in effect identifies Newton's distrust of modern algebraic methods in geometry with a suspicion of the analytical method. On the contrary, Newton acknowledges in so many words, as we have seen, that "in Mathematicks ... the Investigation of difficult Things by the Method of Analysis, ought ever to precede the Method of Composition ...". Newton also practiced what he preached.

His *Universal Arithmetick* was subtitled 'A Treatise of Arithmetical Composition and Resolution'. D. T. Whiteside has gone so far as to say that in the young Newton's mathematical work "a strong bias to the analytical away from the purely geometrical is noticeable".[33] When Newton expresses his disapproval of 'Modern Geometers' who "indulge too much in speculation about Equations", his true reasons are entirely different from a mere suspicion of analysis. They are brought out by a MS fragment recently published. (It is nothing less than a draft of a Preface to the *Principia*.) It shows clearly that Newton was mainly concerned with the lack of stringency in mathematicians' practice of (in effect) presenting analyses without syntheses. Newton writes: "The Mathematicians of the last age have very much improved Analysis but stop there & think they have solved a Problem when they have only resolved it, & by this means the method of Synthesis is almost laid aside."[34]

NOTES

[1] See 'Galileo's Platonism', in M. F. Ashley Montague (ed.), *Studies and Essays in the History of Science and Learning Offered in Homage to George Sarton*, Henry Schuman, New York, 1944, pp. 277–297; *Das Erkenntnisproblem*, Vol. I, Bruno Cassirer, Berlin, 1906, pp. 136–137.

[2] J. H. Randall, Jr., *The School of Padua and the Emergence of Modern Science* (Saggi e Testi), Padua, 1961.

[3] Oskar Becker, *Grösse und Grenze der mathematischen Denkweise*, Verlag Karl Alber, Freiburg and Munich, 1959, pp. 20–25.

[4] G. L. Huxley, 'Two Newtonian Studies', *Harvard Library Bulletin* 13 (1959), 348–361.

[5] See Section (7) below.

[6] Huxley, *op. cit.* (note 4), p. 358.

[7] See especially pp. 304–308 of the original, and cf. pp. 299–300.

[8] *Opticks*, Query 23/31 (1730 edition), Dover reprint, pp. 404–405.

[9] Cf. especially Chapter IV above.

[10] Cf. e.g., Michael S. Mahoney, 'Die Anfänge der algebraischen Denkweise im 17. Jahrhundert', *RETE: Strukturgeschichte der Naturwissenschaften* 1 (1971), 15–31; Michael S. Mahoney, *The Mathematical Career of Pierre de Fermat (1601–65)*, Princeton University Press, Princeton, 1973; and Jacob Klein, *Greek Mathematical Thought and the Origin of Algebra*, The MIT Press, Cambridge, Mass., 1968.

[11] For an interesting parallel attempt by Hooke to hook up the method of analysis with a kind of experimental method, see Mary B. Hesse, 'Hooke's Philosophical Algebra', *Isis* 57 (1966), 67–83, especially 80–83, and further references given there. An especially interesting (though by no means unique) feature of Hooke's ideas is the close connection between analysis and (generalized) algebra. Hooke's ideas differ from Newton's in several respects. It is characteristic that the idea of analysis as analysis of configurations is much less conspicuous in Hooke than in Newton.

[12] Cf. A. C. Crombie, *Medieval and Early Modern Science*, Doubleday, New York,

1959, Vol. II, pp. 11–17, 135–146; A. C. Crombie, *Robert Grosseteste and the Origins of Experimental Science*, Clarendon Press, Oxford, 1953, pp. 27–29, 52–90, 193–194, 297–318.

[13] This seems to have been the case already in Galen; see Neal W. Gilbert, *Renaissance Concepts of Method*, Columbia University Press, New York, 1960, pp. 32–34.

[14] See, e.g., Crombie, *Robert Grosseteste* (note 11 above), pp. 56–57. The quotations given there show that Grosseteste's method of resolution and composition was directional and propositional and did not amount to an analysis of configurations. As a scientific method, it did not have anything to do with the methods of mathematicians. Likewise, the 'anticipations' of Descartes' use of analysis and synthesis quoted by Gilson (in the *Index Scholastico-Cartesien* and in his edition of the *Discours*) all remain within the ambit of the propositional and directional interpretation of analysis.

[15] *Op. cit.*, pp. 172–173.

[16] Crombie, *Robert Grosseteste, loc. cit.*

[17] Henry Guerlac, 'Newton and the Method of Analysis' (preprint), p. 25.

[18] Galileo Galilei, *Dialogue Concerning the Two Chief World Systems*, tr. by Stillman Drake, University of California Press, Berkeley and Los Angeles, 1953, pp. 50–51.

[19] Note also that Galilei in the quoted passage compares in so many words with each other 'the analytical methods' and 'senses, experiments, and observations'. Cf. especially the words 'this is what is done for the most part in the demonstrative sciences'.

[20] Cf. Jaakko Hintikka. 'On the Ingredients of an Aristotelian Science', *Nous* 6 (1972), 55–69.

[21] This point is far from uncontroversial, however. Cf. Galilei, *Dialogues*, Favaro translation, pp. 160–161 (National Edition, p. 197). In any case, Galileo seems to have thought that factual conclusions could be established "by reasoning alone"; see *op. cit.* p. 164 (p. 200 of the National Edition). In one passage (*Opere*, Vol. VII, p. 78) he says that in natural sciences "conclusions are true and necessary, and have nothing to do with human will". Cf. also Thomas P. McTighe, 'Galileo's "Platonism": A Reconsideration', in *Galileo: A Man of Science*, ed. by Ernan McMullin, Basic Books Inc., New York, 1967, pp. 365–387.

[22] The difficulty of understanding Newtons *dicta* otherwise is beautifully demonstrated (it seems to us) by the entirely negative conclusions of Alexandre Koyré's study of them, where the link between hypotheses and the absence of analysis is not heeded. See 'Concept and Experience in Newton's Scientific Thought', Ch. II of Alexandre Koyré, *Newtonian Studies*, Harvard University Press, Cambridge, Mass., 1965.

Out of the enormous body of literature devoted to the subject of Newton's rejection of 'hypotheses', see I. Bernard Cohen, 'Hypotheses in Newton's Philosophy', in *Boston Studies in the Philosophy of Science*, Vol. V, ed. by Robert S. Cohen and Marx W. Wartofsky, D. Reidel, Dordrecht, 1969, pp. 304–326.

[23] *Op. cit.*, pp. 24–25.

[24] Cf. Newton's own formulation (quoted from a MS by Koyré, *op. cit.*, p. 272): "And those things which neither can be demonstrated from the phenomenon nor follow from it by the argument of induction, I hold as hypotheses".

[25] 'Newton and the Inductive Method', in *The* Annus Mirabilis *of Sir Isaac Newton*, ed. by Robert Palter, The MIT Press, Cambridge, Mass., 1970, pp. 244–257.

[26] Guerlac, *op. cit.*, pp. 29–30.

[27] Maurice Mandelbaum, 'Newton and Boyle and the Problem of "Transdiction"', in Maurice Mandelbaum, *Philosophy, Science, and Sense Perception*, The Johns Hopkins Press, Baltimore, 1964.

²⁸ See, e.g., *Discours*, Part II.

²⁹ Cf., e.g., Baillet's account of Descartes' claims for his Method to Cardinal de Bérulle, quoted in N. Kemp Smith, *New Studies in the Philosophy of Descartes*, Macmillan, London, 1952, p. 43.

³⁰ Cf. Guerlac, *op. cit.*, p. 28. The point he makes there can be further generalized.

³¹ Descartes, 'Secundae Responsiones', point 7.

³² For an entry into the enormous literature on this topic, see, e.g., Willis Doney, ed., *Descartes*, Doubleday, Garden City, 1967, pp. 376–377.

³³ D. T. Whiteside, 'Sources and Strengths of Newton's Early Mathematical Thought', in Palter (note 25 above), pp. 69–85, especially p. 75.

³⁴ See I. Bernard Cohen, *Introduction to Newton's Principia*, Cambridge, Cambridge University Press, Cambridge, 1971, pp. 292–294, especially p. 294.

On Newton's use of the method of analysis, see also Zev Bechler, *Newton's Reduction of Optics to Mechanics*, Thesis submitted for the Degree Doctor of Philosophy, Hebrew University, Jerusalem, March, 1972, especially chapter 6, '"Analysis", Induction, Fact and Theory: the Development of Newton's Dialectics'. Cf. also Zev Bechler, 'Newton's Search for a Mechanistic Model of Colour Dispersion: A Suggested Interpretation', *Archive for History of Exact Sciences* **11** (1973), 1–37.

ÁRPÁD K. SZABÓ: WORKING BACKWARDS AND PROVING BY SYNTHESIS

I

The starting point of a historical study on 'Analysis and Synthesis in Greek Science' cannot be but a careful interpretation of the famous passage in the 'Mathematical Collection of Pappus'.[1] So much the more as Hintikka insisted some years ago, and with good reason, on the importance of a distinction between the different senses which the notion analysis and analyticity have had in philosophy. It seems to me, indeed, that exactly the lack of such a distinction had caused much trouble with 'analysis' in ancient science. That is, very often the word itself had not been taken in the same sense as it has in the clear cut explanation of the original Greek text of Pappus. One comes across, e.g., the following usual modern explanations for 'analysis' in Greek mathematics: it could be a process of resolution *of a whole into its parts, into its elements, of the complex into the simple*. Cornford wrote – *inter alia* in lack of mathematical understanding – 'analysis' could be: taking a construction 'to its pieces'.[2] Or one reads sometimes: the Greek expression ἀναλύειν διάγραμμα might mean: "either *to break a figure up into its parts*, or: *to divide a theorem into its premises*".

No doubt, scholars who tried to explain the method of Pappus in such a way, did not catch the sense of the word itself. They confounded the ancient Greek expression (ἀνάλυσις) with its modern counterparts '*analyse*' and '*analysis*'. The exact meaning of these *modern* words runs indeed – as it can easily be checked, e.g., by the aid of *The Concise Oxford Dictionary* – in this way: '*to take apart*', '*taking a compound into its constituents, into its parts*'. Are we, however, to ascribe the same meaning to the original Greek word itself, too? – This seems to me at least very doubtful.

First of all, because the proper Greek expression for '*to take apart*' and '*taking a compound into its constituents*' was not ἀναλύειν and ἀνάλυσις, but διαιρεῖν and διαίρεσις. As it reads in Aristotle, e.g.: "the so-called

elements are last things in which the bodies *can be taken apart*" (διαιρεῖται).[3] Or even in Proclus: "that simpler one in which the compound can
be taken apart."[4] I think this last example is particularly striking, as the
antithetical concepts are: διαιρεῖν and συνθεῖναι exactly for our modern
'*analyse*' and '*synthetize*'. In any case, Greek σύνθεσις and modern
'Synthesis' seem, therefore, to be identical even according to the last
quotation from Proclus, but I don't think that in classical Greek the
words ἀναλύειν and ἀνάλυσις could ever be synonyms for διαιρεῖν and
διαίρεσις respectively. The meaning of the Greek word ἀνάλυσις in everyday language was quite different. You have only to take a dictionary – say
that of Liddell and Scott – in order to see what the most common and usual
meanings of ἀναλύειν and ἀνάλυσις were. That is to say, you will find the
verb ἀναλύειν means '*undo, untie, unloose, relax, unwind*' (namely something which has been *bound* formerly). And therefore the most common
meaning of the noun ἀνάλυσις must be, of course, '*untying, loosening*',
namely '*untying of a "knot", or of a "bend"*'.

Even Pappus does not speak about '*analysis*' as '*taking apart*' or '*taking
a compound into its pieces*'. Instead he says: "And we call such a method
analysis as being a *solution backwards*".[5] That is, according to his explanation the word ἀνάλυσις equals ἀνάπαλιν λύσις: '*untying again, solution
backwards*'. And this is by no means a *pun* on the part of Pappus; quite
the contrary, that is the correct explication of the word itself and of the
mathematical process in question as well. The essence of his account is as
follows.

Analysis is of two kinds; the one is the analysis of the 'problems to
prove' and aims at establishing true theorems – this is the so-called
theoretical kind; the other is the analysis of the 'problems to find' and
aims at finding the unknown – this second one is the so-called *problematical* kind.

In both kinds of analysis we start from what is required, we take it for
granted, and we draw consequences from it, and consequences from the
consequences, till we reach at a *decisive point*. Now I should like to
emphasize that this '*decisive point*' can be, according to the description
by Pappus, in both cases, in the theoretical analysis as well as in the problematical one, *of two different sorts*.

First: if we come, in the process of analysis, upon something *admittedly
false*, that which is sought will also be false (theoretical kind); and

similarly, in the 'problematical' kind: if we come upon something *admittedly impossible*, the problem (the task to solve) will also be impossible. As you see: in the case of a *negative result* the analysis is (in both its forms) *definite*; in such a case there is no reverse operation (σύνθεσις) possible.

The *other possibility* is, however, that we come (e.g. in the *theoretical* analysis) upon a last theorem, about which we have certain knowledge that the theorem is *true*. In such a case our starting theorem will also be true, *provided that all our derivations from the starting theorem are convertible*. Because in that *second case* we must apply the reversing process: in synthesis we start again from the point which we reached last of all in the analysis, from the thing already known or admittedly true. We derive from it what preceded it in the analysis, and go on making derivations until, retracing our steps, we finally succeed in arriving at what is required. This procedure (the second step after analysis) we call synthesis, or constructive solution, or progressive reasoning.

[Now in parenthesis: it is in any case a remarkable feature of the method in question, why the analysis in the case of a positive result – we come in the process of our reasoning upon something admittedly *true* – should be completed with the reverse process, the synthesis? – I should like to cut short this problem in the present connection with two references. – Aristotle said once that, after having made a successful analysis, we sometimes don't succeed in the synthesis. It is not necessary that all our *correct steps* in analysis be also convertible in synthesis.[6] And according to another quotation from him: "If it were impossible to prove *a true conclusion from a false premise*, analysis would be easy, because conclusion and premises would necessarily reciprocate."[7]]

As you see, 'analysis' is, according to the description of Pappus, indeed a *backwards solution*. This is also just why Aristotle, who compared the structure of human deliberation to that of 'analysis', underlined:[8] "The *last* step in analysis is the *first* one in genesis". Putting it in another way: analysis is a reverse solution, a working backwards.

I think it will be very instructive to see how in Antiquity even people who probably had nothing in common with mathematics well knew the fact: 'analysis' in geometry is no 'taking apart', or 'taking a compound into its constituents' but simply: '*working backwards*'. We are told in Plutarch in his 'Vita Romuli' the following amusing though childish story:[9]

It was not known in the days of the Roman antiquarian Varro, in what year and in what month Romulus, the founder of the City had been born. Therefore, Varro gave the task (the problem for solving) to a friend of his, an excellent astrologer to calculate the day and the hour of birth of Romulus, knowing exactly all the important particularities of his biography. The idea is, as it reads in the explanation by Plutarch: if an astrologer gets the exact date (year, month, day and hour) of the birth of somebody, first he makes – with the aid of his tabulations – the constellation for the given moment, and then he can, according to the constellation, forecast the whole future biography of the newborn. Now the astrologer of Varro has to make the *inverse investigation*: knowing the whole biography of his hero, first he must make the constellation for the time of his birth and then fix the exact date when the birth in question took place. – As we are assured by Plutarch, Varro's friend, the competent astrologer could establish in this way not only the exact time of birth of Romulus, but even that of his conception. – Now it is significant that Plutarch compares the method of this astrologer with 'geometrical analysis'; as he says:[10] "from the end out making his reasonings ... as well as the solutions of geometrical problems are being made backwards".

Now before going further and pointing out how the confusion of classical Greek and modern concepts of 'analysis' led to some historical and philosophical misunderstandings, I should like to mention here a rather interesting fact. – Is it possible that all modern scholars lost sight of the big difference between 'analysis' in the modern sense of the word, and the other one which Pappus speaks about? – Not at all! I know at least a few exceptions for contemporary awareness of this difference. First may I quote same words out of the excellent paper by Robinson about our topic. He wrote in 1936:

... when I gave the conventional account of Greek analysis to a mathematical friend he replied that, while he did not see why they called it 'analysis', he himself practised it every day.

I am afraid, these words cannot be understood otherwise than in the sense that while contemporary mathematicians know well and often practise the method of Pappus they don't call it 'analysis', because this same word usually has for them quite a different meaning. This is also just why the mathematician Polya, who gave in his classical works in these last decades a very good interpretation of the method of Pappus, seems to be trying to avoid the word 'analysis' itself. As it reads in his well-known book 'How To Solve It?':[12]

Pappus reports about a branch of study which he calls *analyomenos*. We can render this name in English as 'treasury of analysis', or as 'art of solving problems', or even as 'Heuristic'; the last term seems to be preferable here...

On another occasion Polya calls in the same book[13] the method of 'analysis' (described by Pappus) as the *'method of working backwards'*.

Now I guess, the importance of making clear the ancient meaning of the word 'analysis' consists in the following. If we tried to understand the Greek expression ('analysis') in the modern sense of this same word, we had also to imagine that the geometrical (or mathematical) object 'analysed' in such a way could not be disposed correctly into its organic parts, but always *in one and the same manner*. That is, we had to get in every case necessarily the same organic constituents of the thing taken apart into its very pieces. – This is, however, a misleading idea to my mind when speaking about the 'analysis' of Pappus. Let me remind you: if we make an 'analysis' in the sense of Pappus, we have not simply to deal with the conclusion of a syllogism and we are not seeking plainly for the premises of this same conclusion. The main rule of the celebrated heuristic of Antiquity recorded by Pappus (e.g. for 'problems to prove'), was, as Lakatos puts it:[14] "If you have a conjecture, derive consequences from it, until you arrive at a decisive consequence, which is namely either 'known to be false', or 'known to be true'. At such a decisive consequence the 'analysis' stops in any case, and begins – at least in one of the cases (if you came upon something known to be *true*) – the reverse investigation, i.e. the straightforward 'synthesis'". (I should like to mention here in brackets: I don't think it could make any important difference whether we speak only of *consequences* – in the strict sense of this word – or of *presuppositions* of a tested conjecture as well. The method of Pappus is similarly applicable for both cases!) – Now, the most important preliminary question is for us in the present connection: 'Can *all* consequences of a conjecture (true or false) be thought of as *a single logical chain of statements*? – It seems to me that this would be a mistaken idea, as *almost every conjecture may have many different consequences, that of course don't exclude each other, and that all the same don't make up any close logical chain.* They go only, so to say, together with each other and with the conjecture in question, too. This is why *it depends solely on our own imaginative power which consequences of a given conjecture we discover immediately, and which of them remain hidden for longer from us.* Moreover, I don't believe that even the successful application of the double method of analysis and synthesis in a given case could be full guaranty for having recognized *all* the consequences of the conjecture tested. And this is just why the analytical method does not, its fruitfulness notwithstanding, yield a mechanical effective discovery procedure – as Hintikka puts it.[15] It is just a groping, a fallible procedure which – if it leads to a consequence known to be *true* – does not guarantee the success yet, *at least not without the complementary synthesis.* (Completely different is the case when the 'analysis' leads, so to say, to a 'negative result', if we come in the process of our testing upon something known to be *false.* Such a result is *definite*: our starting theorem was *false!*)

You are also likely to know that there was a major problem in the former literature: whether 'analysis' is to be understood in the text of Pappus as a '*downward movement*' (i.e. as a deduction of consequences from the desired conclusion), or is to be taken as an '*upward movement*' (i.e. as a movement from the desired result to premises from which it could be deduced). And the main question was how these two could coexist in the description of 'analysis' by Pappus. – Now it seems to me that the question posed in this way is only a misunderstanding due to the fact of identifying the 'analysis' of Pappus ('*working backwards*') with modern 'analysing'. Therefore I agree completely with Hintikka when he writes:[16] "There is no problem here because the statements which seem to *present the idea of analysis as a downward movement* are intended to be compatible with the statements which present it as an '*upward path*'". I think the true sense of the Greek scientific term 'analysis' itself is also an argument for this same conception. – On the other hand, being 'analysis' in Greek science a '*working backwards*', I cannot agree, of course, with Hintikka's other formulation, according to which in general 'preoccupation with the *direction of analysis*' should be a sign that the subtler ingredients of the method of analysis ... were overlooked;[17] or even that 'the distinction between *analysis* and *synthesis*' should no longer be 'a difference in direction' but should lie rather in something else.[18] I think 'analysis' is rather merely a testing of a conjecture; as Lakatos puts it:

'a thought-experiment' deploying the original conjecture on a wider front into subconjectures or lemmas, so that our criticism has more targets, while 'synthesis' deals with statements 'known to be true', and constructing with them *proof* for the original conjecture. 'Synthesis' is so to say 'infallible', while 'analysis' is 'fallible'.

II

There is, I think, a very important observation in the paper of Hintikka concerning the exact sense of the Greek logical term ἀκόλουθον in the text of Pappus. As it reads: "We want to suggest that τὸ ἀκόλουθον in Pappus' description of analysis and synthesis *does not mean a logical 'consequence'*, but is a much more vague term for whatever *'corresponds to'* or better *'goes together with'* the desired conclusion in the premises from which can be deduced...".[19] I am fully convinced of the correctness of this observation. Indeed, in general the word ἀκόλουθον does not label any 'logical consequence' in the strict sense of the term, neither in everyday language. I think the next example puts in its proper light the exact meaning of this expression.

We are told in Xenophon's 'Anabasis' the following story.[20] Once in the evening a stranger came to the Greek army in retreat and admonished them: the Greeks had to take care, as their enemies were planning an assault against them in the next night.[21] They had also to keep a watch over the bridge of the next river, because the chief of the enemies had in mind to destroy this bridge.[22] The unexpected news caused alarm among the Greeks, when suddenly a clear-sighted young man observed in the discussion: both messages – the plan of assault, and the other one, that of destroying the bridge – *don't go together, don't harmonize with each other.*[23] – Indeed, why destroy the bridge? Let us suppose the enemies were really planning an assault. In that case first of all they must come over the bridge to the side of the Greeks. Therefore they need the bridge themselves, too. Afterwards, in the second phase, there are two possibilities. Either the enemies would get the upper hand in the clash, and in that case not even an abundance of bridges could be helpful for the defeated Greeks; destroying the bridge would be superfluous from the point of view of the enemies themselves. Or the enemies could even be defeated in the same clash, and in that other case, they wouldn't have – after having previously destroyed the bridge – the possibility of escaping the Greeks: destroying the bridge could be dangerous even for the enemies themselves.

As you see, both statements – the 'plan of assault', and 'that of destroying the bridge' – are independent of one another; both of them could be true in themselves. In the given situation, however, *they don't go with each other*, as the Greek say: they are οὐκ ἀκόλουθα. – But let us now imagine the opposite of the case just described. Even if both facts don't exclude each other, and could thus go together, even then none of them could be thought of as a 'consequence' of the other. – Therefore it seems to me that Hintikka hit the nail on the head, pointing out the somewhat *uncertain* meaning of the terms ἀκόλουθα and οὐκ ἀκόλουθα respectively. That is: ἀκόλουθα are called the things which *go together*, without having any closer logical link between themselves. Indeed, in the analysis of Pappus we are not looking properly speaking for the which 'logical consequences' of the thing sought, which we had supposed to be true at the beginning of our process of analysis; instead we are only groping for what would go together with the same statement, if it would be true. – And this is just one of the most important reasons why 'analysis' in the sense of Pappus is not 'taking apart a compound into its constituents'. When we are making such an 'analysis' we are not looking for the very pieces of a whole, instead we are trying to find what could harmonize with our proposition in question, if it happened to be true. And this is *no modern 'analysing'*.

On the other hand, I am more sceptical as regards the terminological distinction of Hintikka between οὐκ ἀκόλουθον and οὐ ἕπεται. This last one should namely express – as Hintikka tried to point out – in contrast to the other one the real *'lack of a deductive connection'*.[24] Frankly speaking, I don't think that such a distinction is more than a modern over-interpretation of the ancient texts. To my mind all expressions of that sort, as τὰ ἀκόλουθα, τὰ ἑπόμενα, τὰ ἕξης, τὰ συμβαίνοντα, τὰ συμφωνήεντα, etc. are only synonyms, derived as technical terms from the old dialectical tradition of the eleatic-platonic philosophy, as I tried to point out recently in my book.[25] That is to say, if these designations come to be used in a *positive* sense, they express that there is *no contradiction* between two or more statements in question, while if they are used in a *negative* sense, they always stress some sort of contradiction, namely that the statements in question *don't go together with*. Therefore I think the tech-

nical terms just mentioned are, among others, the evidence for the origin
of the method of 'analysis and synthesis' out of the dialectic.

In that sense I should like to complete Hintikka's correct observation
concerning the logical term ἀκόλουθον with two important statements.

First: the logical expression ἀκόλουθον (or in its negative form: οὐκ
ἀκόλουθον) is a well-known term of the so-called *reductio ad absurdum* =
= proof leading to the impossible. (By the way: I have dealt with this
very important form of mathematical proof and its technical terms in my
book: 'Anfänge, etc.')

Second: the 'analysis' of Pappus that leads to something known being
false is just a special case of the so-called *reductio ad absurdum*; because in
such a case our starting theorem is, beyond any doubt, *false*. (Putting it
in another way: the contrary of the same starting theorem must be true.)

III

Let me return now to Pappus' text. (I use again Lakatos' summary.)

Draw conclusions from your conjecture, one after the other, assuming that it is true.
If you reach a false conclusion, then your conjecture was false. (This is a particularly
important point! As we have seen, the 'analysis' with such a negative result is a 'reductio
ad absurdum'.) If you reach, on the other hand, an indubitably true conclusion, your
conjecture may have been true. (Please, note! If you reach a true conclusion in the
process of your analysis, there is no guaranty yet for the original conjecture being true
in any case! It is only a favourable auspice: also the original conjecture *may have been
true*!) Therefore, in that latter case, reverse now the process as *synthesis*, and try to
deduce your original conjecture through the inverse route from the indubitable truth
to the dubitable conjecture. If you succeed, you have proved your original conjecture.[26]

Let us now have a closer look at the *synthesis*. As Hankel stated it,[27]
the proofs for mathematical theorems were made in Antiquity generally
by *synthesis* (except the cases when the proof was leading to the impossible,
reductio ad absurdum). Of course, we must not confound the proof by

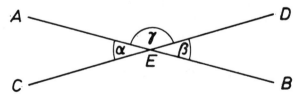

Fig. 3.

synthesis with a simple syllogism. A proof by synthesis is, e.g., the following in the case of Proposition 15 in Book I of Euclid's 'Elements'. This proposition says:

> *"If two straight lines cut one another, they make the vertical angles equal to one another."*

That is to say: in our case (see Figure 3) the two angles α and β are equal (α = β). We prove this statement by synthesis in the following *four* steps.

(1) We look at the straight lines *CD*, *AE*, and we see that the sum of the two angles α + γ must make two rights. Then we know the Proposition I.13: *"If a straight line set up on a straight line make two angles, it will make either two right angles or angles equal to two right angles"* – and this is just our case. Therefore: α + γ = 2R. – You see, the first step in the synthesis is the application of a well known theorem to the case in question.

(2) The second step is again the application of the same theorem to the case in question *from another point of view*. Then we look at the straight lines *AB*, *ED*, and we see that the sum of the other two angles γ + β must also be equal to two right angles, γ + γ = 2R.

(3) Now we have – after the two first steps – the following two equations: α + γ = 2R and α + β = 2R. Therefore we can make our 3rd step, which is the application of *Axiom* 1: *"Things which are equal to the same thing are also equal to one another"*. Therefore in our case: α + γ = γ + β.

(4) And at last the 4th step will be the application of *Axiom* 3: *"If equals be subtracted from equals, the remainders are equal"*. In our case: α + (not) γ = (not) γ + β, and that is the conclusion of our proving by synthesis.

As you see, the single steps are not statements that could be called 'consequences' or 'presuppositions' of each other. They *go only together with*, that is: they *harmonize with each other*, and therefore they can make up a proof.

Let us now reverse the same synthesis: How could it be prepared by a foregoing *analysis* in the sense of Pappus?

(1) The first step of the 'analysis' in the present case could be only that *we suppose the thing sought as being true*: α = β.

(2) In the second step we say: we don't know in fact whether our

theorem is true, but if it is true, we could apply, e.g., *Axiom* 2: "*If equals be added to equals, the wholes are equal*". And so we could get: $\alpha + \gamma = \beta + \gamma$.

Please note, in that connection the following. It is not at all necessary that our second step in the analysis be just that. It could be something else as well! Analysis is in the sense of Pappus nothing else but a groping, a tentative! Besides, this *second step* of our analysis does not play any role in the synthesis. *Analysis* and *synthesis* may complete each other mutually, but they need not consist exactly of the same steps!

(3) Now in step 3 we ask: Is our last equation true? What do we see, e.g., on the left side of the equation? $\alpha + \gamma$: according to the figure these two angles must be equal to two right angles (Proposition I.13). But the same is also valid for the right side of our equation. We see, therefore, that the equation is true. – We have come upon a statement known to be true, and so we can now try to reverse our analysis in a synthesis, as it has been pointed out beforehand.

At the end of this modest contribution I should like to emphasize the following points.

(1) It seems to me that the method of Pappus – 'analysis and synthesis' – is nothing but a further developed procedure of the so-called 'proof leading to the impossible'. I think, this statement is – at least in the case when 'analysis' leads to something *known to be false* – obvious. – On the other hand, if we come in our process of 'analysis' to a statement *known to be true*, we have to complete our 'analysis' with the reverse process ('synthesis'), *because a true conclusion from a false premise* is never excluded.

(2) The modern investigation of this ancient method began last century with Hankel's excellent book. For further insights we are mostly indebted, among recent scholars, to Polya (see his works in the References), and above all to Lakatos, who tried, even in his 'Proofs and Refutations' to improve the method of Pappus.

NOTES

[1] Pappus, *Collectio Mathematica*, ed. by F. Hultsch, Berlin 1876–1878, pp. 634–636.
[2] F. M. Cornford, p. 46.
[3] Aristotle, *Met.* 1014a31: τὰ τῶν σωμάτων στοιχεῖα λέγουσιν ... εἰς ἃ διαιρεῖται τὰ σώματα ἔσχατα.

4 'Proclus Diadochus', in *Euclidis*, elem. librum primum comm., ed. by G. Friedlein, Lips. 1873 (in Eucl. I. 72: εἰς ὃ ἀπλούστερον διαιρεῖται τὸ σύνθετον.

5 II. 634:καὶ τὴν τοιαύτην ἔφοδον ἀνάλυσιν καλοῦμεν οἷον ἀνάπαλιν λύσιν.

6 Aristotle, *Top.* 175a27: συμβαίνει δέ ποτε καθάπερ ἐν ταῖς διαγράμμασιν, καὶ γὰρ ἐκεῖ ἀναλύσαντες ἐνίοτε συνθεῖναι πάλιν ἀδυνατοῦμεν.

7 Aristotle, *An. Post.* I, 12, 78a7 sqq. (Quotation above according to the text given earlier in the present volume, p. 20, n. 5.)

8 *Eth. Nic.* 1112b15–25: τὸ ἔσχατον ἐν τῇ ἀναλύσει πρῶτον ἐν τῇ γενέσει.

9 Cap. 12.

10 *Ibid.*: ἐκ τῶν ἀποτελεσμάτων ... ποιησάμενον τὸν συλλογισμόν, ὥσπερ αἱ τῶν γεωμετρικῶν ὑφηγοῦνται προβλημάτων ἀναλύσεις.)

11 R. Robinson, p. 469.

12 P. 141.

13 *Ibid.*, p. 225.

14 'Proofs and Refutations', p. 68 n.

15 This volume, p. 48.

16 Above, p. 13.

17 Above, p. 11.

18 Above, p. 44.

19 Above, p. 14.

20 Xenophon, *Exp. Cyri* (L. Dindorf, 1875), lib. II, Cap. IV.

21 μὴ ὑμῖν ἐπιθῶνται τῆς νυκτὸς οἱ βάρβαροι.

22 ὡς διανοεῖται αὐτὴν (scil. τὴν γέφυραν) λῦσαι ... τῆς νυκτός.

23 οὐκ ἀκόλουθα εἴη τό τε ἐπιθήσεσθαι καὶ λύσειν τὴν γέφυραν.

24 Above, p. 15.

25 'Anfänge etc.', p. 315, 321, 326.

26 Lakatos (II), p. 197.

27 H. Hankel, p. 240.

BIBLIOGRAPHY

Cornford, F. M., 'Mathematics and Dialectic in the Republic VI–VII', *Mind* **41** (1932), 37–52; 173–190.

Einarson, B., *American Journal of Philology* **57** (1936) 33–54; 151–172.

Gulley, N., 'Greek Geometrical Analysis', *Phronesis* **33** (1958), 1–14.

Hankel, H., *Zur Geschichte der Mathematik*, Leipzig, 1874.

Heath, T. L., (I) *Euclid's Elements*, Vol. I–III, Dover Publications.

Heath, T. L., (II) *Mathematics in Aristotle*, Oxford, 1949.

Hintikka, K. J. J., 'An Analysis of Analyticity', in Paul Weingartner (ed.), *Deskription, Analytizität und Existenz*, Pustet, Salzburg and Munich, 1966, pp. 193–214.

Lakatos, I., 'Proofs and Refutations', *The British Journal for the Philosophy of Science* **14** (1963–64), Nos. 53–56, Thomas Nelson and Sons Ltd., Edinburgh.

Lakatos, I., 'Essay in the Logic of Mathematical Discovery', University of Cambridge, Diss. 1961, pp. 193–242.

Mahoney, S. Michael, 'Another Look at Greek Geometrical Analysis', *Archive for History of Exact Sciences* **5** (1968), 318–348.

Polya, G., *How to Solve It?*, second edition, Doubleday Anchor Books, Garden City–New York, 1957.

Polya, G., *Induction and Analogy*, Vol. I–II, Princeton University Press, Princeton, 1954.

Polya, G., *Mathematical Discovery*, Vol. I–II, John Wiley and Sons Inc., New York–London, 1962, 1965.

Robinson, R., 'Analysis in Greek Geometry', *Mind* 45 (1936), 464–473.

Szabó, A., *Anfänge der griechischen Mathematik*, München-Wien 1969.

Szabó, A., 'Der math. Begriff *dynamis*', *Maia N.S.* 15 (1963), 219–256.

Waerden, B. L. v. d., *Erwachende Wissenschaft*, Basel-Stuttgart, 1956.

Animalium I, 18, 724b28 and *De Mundo* 4, 394b17. As far as mathematical analysis is concerned, we have Proclus' testimony (see Friedlein, p. 57). He speaks of the geometrical analysis (as opposed to synthesis) as a transition from *complex* matters into more *simple* ones. This passage shows that Prof. Szabó's claim does not find real support in Proclus, either.

However, even if Professor Szabó's thesis about *analysis* and *analyein* as 'untying' were correct, it would not be conclusive for the substantial issues. For one thing, our arguments for the analysis-of-figures view were not based in the least on the meaning of the ancient Greek terms *analysis* and *analyein*. The customary view of their meaning was used only to *illustrate* our thesis, not to *argue for* it, although a closer analysis of their import can probably serve to supply further evidence for our view.

Furthermore, Szabó's interpretation of analysis as 'untying' or 'unwinding' or 'solution backwards' does not have any specific implications concerning the direction of analysis or its heuristic significance. The idea of 'backwards solution' encourages, at least *prima facie*, the idea of analysis as proceeding 'upwards', against the direction of relations of logical consequence. Yet Szabó opts for the usual interpretation according to which we are in an analysis moving 'downwards' and only able to climb back synthetically in virtue of the convertibility of all the steps of analysis. We have seen (cf. especially Chapter IV above) that although this view contains a great deal of truth, its justification is only seen when the structure of analysis is analysed in greater detail than Szabó does. An indication of his problems in this department is his assumption that an analysis ends up with a *theorem*, true or false. (Cf. p. 120, lines 7–8, of Szabó's paper.) This is simply not the case in any actual example of (theoretical) analysis that can easily be found in the Greek mathematical corpus. Rather, an analysis (conceived of as a 'downward' movement, as Szabó does) typically ends with a statement concerning the relationship of the ingredients of a geometrical configuration which is implied by earlier theorems in conjunction with the 'given' antecedent assumptions and the hypothetical *zetoumenon*-assumption concerning the configuration, or else with a negation of one. The antecedent assumptions (our '*A*' on pp. 36 and 41) are what we argued Pappus' *arkhai* to be at Hultsch 634, line 17.

We are somewhat puzzled by Szabó's remarks on the modern sense of analysis in connection with contemporary mathematicians like our delightful Pappus *redivivus*, G. Polya, for the awkward modern sense of 'analysis' which has prevented them from so labelling Pappus' method is of course not 'taking apart' but 'analysis' in the sense of the familiar label 'higher' or 'infinitesimal analysis' for a certain mathematical discipline.

The most important question that arises here concerns of course the *rationale* of the heuristic significance of the analytical method. Szabó in effect objects to the analysis-of-figures interpretation because on that view analysis would (in his view) always proceed *"in one and the same manner"* (p. 122 of Szabó's paper; his italics). Presumably analysis would according to Szabó's interpretation of the analysis-of-figures view come to an end automatically after a finite number of steps.

Here we are tempted to say: so much the better for the analysis-of-figures view. If what Szabó says of the analysis-of-figures idea were the case, analysis would indeed be a marvellous heuristic method, so good a method of invention that we might be reluctant to call it a merely heuristic procedure, since it would amount to a foolproof discovery method. It is *prima facie* very strange to see Professor Szabó rejecting the analysis-of-figures idea on the ground that explains so very well the heuristic significance of the analytical method. We believe that he is here in fact touching on the sources of the heuristic usefulness of the use of analysis, and hence pointing out a reason for preferring the analysis-of-figures view rather than rejecting it.

Of course Szabó's instincts are sound when he calls the idea of a unique way of (as it were) disassembling a given figure in analysis 'misleading' (p. 122, line 16). It is a brute fact every discussion of the heuristics of mathematics has to recognize that no such uniqueness and consequently no foolproof discovery procedure is in general possible. But Szabó is in our opinion mistaken when he uses this fact to dismiss the analysis-of-figures interpretation in favour of the idea of analysis as a method of "testing conjectures".

First, the failure of the analytical method as a foolproof discovery procedure has to be explained by reference to the need of constructions (auxiliary individuals). It need not be explained in the way Szabó does. Analysis would indeed yield a discovery procedure if no *kataskeue* were needed. There is indeed something unique and closed in the analysis of a given figure. However, an analyst usually needs auxiliary constructions, and sight unseen he often cannot be ever quite sure that he has carried out enough of them. What he must analyse is not a unique figure determined completely by the theorem or problem at hand, but a figure amplified by constructions in the finding of which (as Leibniz already saw – see *Nouveaux Essais* IV, xvii, 3) "the greatest art" of an analyst typically consists.

For this reason, Szabó's beautiful little sample analysis (pp. 126–8) is misleading, since no auxiliary constructions are needed in it. We regret that he has not chosen to bring his unique expertise in Greek mathematics to bear on this central problem of the role of constructions in geometrical (and other mathematical) arguments.

Secondly, the idea of analysis as a "testing of conjectures", while not incompatible with our interpretation, does not contribute much to a genuine theoretical understanding of the heuristic usefulness of the analytical method. This charge is of course not directed specifically to Szabó. The interpretation of the old method of analysis is an important challenge to historians and philosophers alike – and to their joint efforts in particular – just because of the surprising fact that *no rational explanation of the usefulness of the method is to be found in the literature.* Anyone who doubts this is welcome to examine the literature. For instance, references to the direction of analysis are useless here, for *a priori* this direction should make no difference to the ease or difficulty of a mathematician's journey from old theorems to a new one. (See the last paragraph on p. 18 of this book. It is only fair to mention that this paragraph was added after Szabó received his version of our paper but before his comments reached us.)[1] Not only is it the case that Szabó's attempted explanation of the usefulness of analysis is not needed. What is more important, it does not explain the usefulness as it stands.

The development of the method of analysis from the idea of a proof by *reductio ad absurdum* is an interesting and highly plausible historical thesis, but it does not serve to explain the usefulness of either procedure. The remarkable success of Polya's heuristic remarks on what is a descendant of the ancient method of analysis is of course due more to his amazing pedagogical skills than to any rational analysis of the logic of the analytical method.

As to the meaning of τὰ ἀκόλουθα, τὰ ἑπόμενα, τὰ ἑξῆς, τὰ συμβαίνοντα, etc., a couple of remarks are in order. First, we did not make any claims concerning their meaning in Greek mathematics at large. What we did argue was that for one writer, Pappus, τὰ ἀκόλουθα and τὰ ἑπόμενα were not synonyms but were rather used in markedly different ways, and that for him τὰ ἀκόλουθα did not seem to bear any sharp technical sense. We do not see that the interesting evidence Szabó refers to is relevant to these specific theses.

Furthermore, it seems to us misleading to say, as Szabó says of his sample analysis, that the successive steps of synthesis are not logical consequences. Of course they are not logical consequences of earlier steps alone. This is trivial, and it does not affect the fact that each step is in fact a logical consequence of earlier steps, earlier theorems, and axioms, *together with* the given data describing the geometrical configuration in question. In terms of our analysis of analysis, the steps are not logical consequences of K alone, but of K in conjunction with A. Who's afraid of big bad logic here?

NOTE

[1] It is also clear that the multiplicity of consequences which any conjecture has and which is emphasized by Szabó does not suffice to explain why we have to resort to heuristics (to "our own imaginative power", as Szabó puts it) nor why the particular way of going about which is utilized in analysis is heuristically valuable. Of course most propositions have many consequences, but *vice versa* there usually also are many others from which it could be inferred.

INDEX OF NAMES

INDEX OF SUBJECTS

INDEX OF PASSAGES

SYNTHESE LIBRARY

Monographs on Epistemology, Logic, Methodology,
Philosophy of Science, Sociology of Science and of Knowledge, and on the
Mathematical Methods of Social and Behavioral Sciences

Editors:

DONALD DAVIDSON (The Rockefeller University and Princeton University)
JAAKKO HINTIKKA (Academy of Finland and Stanford University)
GABRIËL NUCHELMANS (University of Leyden)
WESLEY C. SALMON (University of Arizona)

1. J. M. BOCHEŃSKI, *A Precis of Mathematical Logic.* 1959, X+100 pp.
2. P. L. GUIRAUD, *Problèmes et méthodes de la statistique linguistique.* 1960, VI+146 pp.
3. HANS FREUDENTHAL (ed.), *The Concept and the Role of the Model in Mathematics and Natural and Social Sciences, Proceedings of a Colloquium held at Utrecht, The Netherlands, January 1960.* 1961, VI+194 pp.
4. EVERT W. BETH, *Formal Methods. An Introduction to Symbolic Logic and the Study of Effective Operations in Arithmetic and Logic.* 1962, XIV+170 pp.
5. B. H. KAZEMIER and D. VUYSJE (eds.), *Logic and Language. Studies dedicated to Professor Rudolf Carnap on the Occasion of his Seventieth Birthday.* 1962, VI+256 pp.
6. MARX W. WARTOFSKY (ed.), *Proceedings of the Boston Colloquium for the Philosophy of Science, 1961–1962,* Boston Studies in the Philosophy of Science (ed. by Robert S. Cohen and Marx W. Wartofsky), Volume I. 1973, VIII+212 pp.
7. A. A. ZINOV'EV, *Philosophical Problems of Many-Valued Logic.* 1963, XIV+155 pp.
8. GEORGES GURVITCH, *The Spectrum of Social Time.* 1964, XXVI+152 pp.
9. PAUL LORENZEN, *Formal Logic.* 1965, VIII+123 pp.
10. ROBERT S. COHEN and MARX W. WARTOFSKY (eds.), *In Honor of Philipp Frank,* Boston Studies in the Philosophy of Science (ed. by Robert S. Cohen and Marx W. Wartofsky), Volume II. 1965, XXXIV+475 pp.
11. EVERT W. BETH, *Mathematical Thought. An Introduction to the Philosophy of Mathematics.* 1965, XII+208 pp.
12. EVERT W. BETH and JEAN PIAGET, *Mathematical Epistemology and Psychology.* 1966, XII+326 pp.
13. GUIDO KÜNG, *Ontology and the Logistic Analysis of Language. An Enquiry into the Contemporary Views on Universals.* 1967, XI+210 pp.
14. ROBERT S. COHEN and MARX W. WARTOFSKY (eds.), *Proceedings of the Boston Colloquium for the Philosophy of Science 1964–1966, in Memory of Norwood Russell Hanson,* Boston Studies in the Philosophy of Science (ed. by Robert S. Cohen and Marx W. Wartofsky), Volume III. 1967, XLIX+489 pp.
15. C. D. BROAD, *Induction, Probability, and Causation. Selected Papers.* 1968, XI+296 pp.
16. GÜNTHER PATZIG, *Aristotle's Theory of the Syllogism. A Logical-Philosophical Study of Book A of the Prior Analytics.* 1968, XVII+215 pp.

17. NICHOLAS RESCHER, *Topics in Philosophical Logic*. 1968, XIV+347 pp.
18. ROBERT S. COHEN and MARX W. WARTOFSKY (eds.), *Proceedings of the Boston Colloquium for the Philosophy of Science 1966–1968*, Boston Studies in the Philosophy of Science (ed. by Robert S. Cohen and Marx W. Wartofsky), Volume IV. 1969, VIII+537 pp.
19. ROBERT S. COHEN and MARX W. WARTOFSKY (eds.), *Proceedings of the Boston Colloquium for the Philosophy of Science 1966–1968*, Boston Studies in the Philosophy of Science (ed. by Robert S. Cohen and Marx W. Wartofsky), Volume V. 1969, VIII+482 pp.
20. J. W. DAVIS, D. J. HOCKNEY, and W. K. WILSON (eds.), *Philosophical Logic*. 1969, VIII+277 pp.
21. D. DAVIDSON and J. HINTIKKA (eds.), *Words and Objections: Essays on the Work of W. V. Quine*. 1969, VIII+366 pp.
22. PATRICK SUPPES, *Studies in the Methodology and Foundations of Science. Selected. Papers from 1911 to 1969*, XII+473 pp.
23. JAAKKO HINTIKKA, *Models for Modalities. Selected Essays*. 1969, IX+220 pp.
24. NICHOLAS RESCHER *et al.* (eds.). *Essay in Honor of Carl G. Hempel. A Tribute on the Occasion of his Sixty-Fifth Birthday*. 1969, VII+272 pp.
25. P. V. TAVANEC (ed.), *Problems of the Logic of Scientific Knowledge*. 1969, XII+429 pp.
26. MARSHALL SWAIN (ed.), *Induction, Acceptance, and Rational Belief*. 1970. VII+232 pp.
27. ROBERT S. COHEN and RAYMOND J. SEEGER (eds.), *Ernst Mach: Physicist and Philosopher*, Boston Studies in the Philosophy of Science (ed. by Robert S. Cohen and Marx W. Wartofsky), Volume VI. 1970, VIII+295 pp.
28. JAAKKO HINTIKKA and PATRICK SUPPES, *Information and Inference*. 1970, X+336 pp.
29. KAREL LAMBERT, *Philosophical Problems in Logic. Some Recent Developments*. 1970, VII+176 pp.
30. ROLF A. EBERLE, *Nominalistic Systems*. 1970, IX+217 pp.
31. PAUL WEINGARTNER and GERHARD ZECHA (eds.), *Induction, Physics, and Ethics, Proceedings and Discussions of the 1968 Salzburg Colloquium in the Philosophy of Science*. 1970, X+382 pp.
32. EVERT W. BETH, *Aspects of Modern Logic*. 1970, XI+176 pp.
33. RISTO HILPINEN (ed.), *Deontic Logic: Introductory and Systematic Readings*. 1971, VII+182 pp.
34. JEAN-LOUIS KRIVINE, *Introduction to Axiomatic Set Theory*. 1971, VII+98 pp.
35. JOSEPH D. SNEED, *The Logical Structure of Mathematical Physics*. 1971, XV+311 pp.
36. CARL R. KORDIG, *The Justification of Scientific Change*. 1971, XIV+119 pp.
37. MILIČ ČAPEK, *Bergson and Modern Physics*, Boston Studies in the Philosophy of Science (ed. by Robert S. Cohen and Marx W. Wartofsky), Volume VII. 1971, XV+414 pp.
38. NORWOOD RUSSELL HANSON, *What I do not Believe, and other Essays*, ed. by Stephen Toulmin and Harry Woolf. 1971, XII+390 pp.
39. ROGER C. BUCK and ROBERT S. COHEN (eds.), *PSA 1970. In Memory of Rudolf Carnap*, Boston Studies in the Philosophy of Science (ed. by Robert S. Cohen and Marx W. Wartofsky), Volume VIII. 1971, LXVI+615 pp. Also available as a paperback.

40. DONALD DAVIDSON and GILBERT HARMAN (eds.), *Semantics of Natural Language.* 1972, X+769 pp. Also available as a paperback.
41. YEHOSUA BAR-HILLEL (ed.), *Pragmatics of Natural Languages.* 1971, VII+231 pp.
42. SÖREN STENLUND, *Combinators, λ-Terms and Proof Theory.* 1972, 184 pp.
43. MARTIN STRAUSS, *Modern Physics and Its Philosophy. Selected Papers in the Logic, History, and Philosophy of Science.* 1972, X+297 pp.
44. MARIO BUNGE, *Method, Model and Matter.* 1973, VII+196 pp.
45. MARIO BUNGE, *Philosophy of Physics.* 1973, IX+248 pp.
46. A. A. ZINOV'EV, *Foundations of the Logical Theory of Scientific Knowledge (Complex Logic)*, Boston Studies in the Philosophy of Science (ed. by Robert S. Cohen and Marx W. Wartofsky), Volume IX. Revised and enlarged English edition with an appendix, by G. A. Smirnov, E. A. Sidorenka, A. M. Fedina, and L. A. Bobrova 1973, XXII+301 pp. Also available as a paperback.
47. LADISLAV TONDL, *Scientific Procedures*, Boston Studies in the Philosophy of Science (ed. by Robert S. Cohen and Marx W. Wartofsky), Volume X. 1973, XII+268 pp. Also available as a paperback.
48. NORWOOD RUSSELL HANSON, *Constellations and Conjectures*, ed. by Willard C. Humphreys, Jr. 1973, X+282 pp.
49. K. J. J. HINTIKKA, J. M. E. MORAVCSIK, and P. SUPPES (eds.), *Approaches to Natural Language. Proceedings of the 1970 Stanford Workshop on Grammar and Semantics.* 1973, VIII+526 pp. Also available as a paperback.
50. MARIO BUNGE (ed.), *Exact Philosophy – Problems, Tools, and Goals.* 1973, X+214 pp.
51. RADU J. BOGDAN and ILKKA NIINILUOTO (eds.), *Logic, Language, and Probability.* A selection of papers contributed to Sections IV, VI, and XI of the Fourth International Congress for Logic, Methodology, and Philosophy of Science, Bucharest, September 1971. 1973, X+323 pp.
52. GLENN PEARCE and PATRICK MAYNARD (eds.), *Conceptual Chance.* 1973, XII+282 pp.
53. ILKKA NIINILUOTO and RAIMO TUOMELA, *Theoretical Concepts and Hypothetico-Inductive Inference.* 1973, VII+264 pp.
54. ROLAND FRAÏSSÉ, *Course of Mathematical Logic – Volume I: Relation and Logical Formula.* 1973, XVI+186 pp. Also available as a paperback.
55. ADOLF GRÜNBAUM, *Philosophical Problems of Space and Time.* Second, enlarged edition, Boston Studies in the Philosophy of Science (ed. by Robert S. Cohen and Marx W. Wartofsky), Volume XII. 1973, XXIII+884 pp. Also available as a paperback.
56. PATRICK SUPPES (ed.), *Space, Time, and Geometry.* 1973, XI+424 pp.
57. HANS KELSEN, *Essays in Legal and Moral Philosophy*, selected and introduced by Ota Weinberger. 1973, XXVIII+300 pp.
58. R. J. SEEGER and ROBERT S. COHEN (eds.), *Philosophical Foundations of Science. Proceedings of an AAAS Program, 1969.* Boston Studies in the Philosophy of Science (ed. by Robert S. Cohen and Marx W. Wartofsky), Volume XI. 1974, X+545 pp. Also available as paperback.
59. ROBERT S. COHEN and MARX W. WARTOFSKY (eds.), *Logical and Epistemological Studies in Contemporary Physics*, Boston Studies in the Philosophy of Science (ed. by Robert S. Cohen and Marx W. Wartofsky), Volume XIII. 1973, VIII+462 pp. Also available as paperback.
60. ROBERT S. COHEN and MARX W. WARTOFSKY (eds.), *Methodological and Historical*

Essays in the Natural and Social Sciences. Proceedings of the Boston Colloquium for the Philosophy of Science, 1969–1972, Boston Studies in the Philosophy of Science (ed. by Robert S. Cohen and Marx. W. Wartofsky), Volume XIV. 1974, VIII+405 pp. Also available as paperback.

61. ROBERT S. COHEN, J. J. STACHEL, and MARX W. WARTOFSKY (eds.), *For Dirk Struik. Scientific, Historical and Political Essays in Honor of Dirk J. Struik,* Boston Studies in the Philosophy of Science (ed. by Robert S. Cohen and Marx W. Wartofsky), Volume XV. 1974, XXVII+652 pp. Also available as paperback.

62. KAZIMIERZ AJDUKIEWICZ, *Pragmatic Logic,* transl. from the Polish by Olgierd Wojtasiewicz. 1974, XV+460 pp.

63. SÖREN STENLUND (ed.), *Logical Theory and Semantic Analysis. Essays Dedicated to Stig Kanger on His Fiftieth Birthday.* 1974, V+217 pp.

64. KENNETH F. SCHAFFNER and ROBERT S. COHEN (eds.), *Proceedings of the 1972 Biennial Meeting, Philosophy of Science Association,* Boston Studies in the Philosophy of Science (ed. by Robert S. Cohen and Marx W. Wartofsky), Volume XX. 1974, IX+444 pp. Also available as paperback.

65. HENRY E. KYBURG, JR., *The Logical Foundations of Statistical Inference.* 1974, IX+421 pp.

66. MARJORIE GRENE, *The Understanding of Nature: Essays in the Philosophy of Biology,* Boston Studies in the Philosophy of Science (ed. by Robert S. Cohen and Marx W. Wartofsky), Volume XXIII. 1974, XII+360 pp. Also available as paperback.

67. JAN M. BROEKMAN, *Structuralism: Moscow, Prague, Paris.*

68. NORMAN GESCHWIND, *Selected Papers on Language and the Brain,* Boston Studies in the Philosophy of Science (ed. by Robert S. Cohen and Marx W. Wartofsky), Volume XVI. 1974, XII+549 pp. Also available as paperback.

69. ROLAND FRAÏSSÉ. *Course of Mathematical Logic – Volume II: Model Theory.* 1974, XIX+192 pp.

70. ANDRZEJ GRZEGORCZYK, *An Outline of Mathematical Logic. Fundamental Results and Notions Explained with All Details.* 1974, X+596 pp.

SYNTHESE HISTORICAL LIBRARY

Texts and Studies
in the History of Logic and Philosophy

Editors:

N. KRETZMANN (Cornell University)
G. NUCHELMANS (University of Leyden)
L. M. DE RIJK (University of Leyden)

1. M. T. BEONIO-BROCCHIERI FUMAGALLI, *The Logic of Abelard*. Translated from the Italian. 1969, IX+101 pp.

2. GOTTFRIED WILHELM LEIBNITZ, *Philosophical Papers and Letters*. A selection translated and edited, with an introduction, by Leroy E. Loemker. 1969, XII+736 pp.

3. ERNST MALLY, *Logische Schriften*, ed. by Karl Wolf and Paul Weingartner. 1971, X+340 pp.

4. LEWIS WHITE BECK (ed.), *Proceedings of the Third International Kant Congress*. 1972, XI+718 pp.

5. BERNARD BOLZANO, *Theory of Science*, ed. by Jan Berg. 1973, XV+398 pp.

6. J. M. E. MORAVCSIK (ed.), *Patterns in Plato's Thought. Papers arising out of the 1971 West Coast Greek Philosophy Conference*. 1973, VIII+212 pp.

7. NABIL SHEHABY, *The Propositional Logic of Avicenna: A Translation from al-Shifā':al-Qiyās*, with Introduction, Commentary and Glossary. 1973, XIII+296 pp.

8. DESMOND PAUL HENRY, *Commentary on De Grammatico: The Historical-Logical Dimensions of a Dialogue of St. Anselm's*. 1974, IX+345 pp.

9. JOHN CORCORAN, *Ancient Logic and Its Modern Interpretations*. 1974. X+208 pp.

10. E. M. BARTH, *The Logic of the Articles in Traditional Philosophy*. 1974, XXVII+533 pp.

11. JAAKKO HINTIKKA, *Knowledge and the Known. Historical Perspectives in Epistemology*. 1974, XII+243 pp.

12. E. J. ASHWORTH, *Language and Logic in the Post-Medieval Period*. 1974, XIII+304 pp.